Evolving Enactivism

Evolving Enactivism

Basic Minds Meet Content

Daniel D. Hutto and Erik Myin

The MIT Press
Cambridge, Massachusetts
London, England

© 2017 Massachusetts Institute of Technology

All rights reserved. No part of this book may be reproduced in any form by any electronic or mechanical means (including photocopying, recording, or information storage and retrieval) without permission in writing from the publisher.

This book was set in ITC Stone Sans Std and ITC Stone Serif Std by Toppan Best-set Premedia Limited. Printed and bound in the United States of America.

Library of Congress Cataloging-in-Publication Data

Names: Hutto, Daniel D., author. | Myin, Erik, author.
Title: Evolving enactivism : basic minds meet content / Daniel D. Hutto and Erik Myin.
Description: Cambridge, MA : MIT Press, [2017] | Includes bibliographical references and index.
Identifiers: LCCN 2016039862 | ISBN 9780262036115 (hardcover : alk. paper)
Subjects: LCSH: Philosophy of mind. | Cognitive science. | Act (Philosophy) | Intentionalism. | Mental representation. | Intentionality (Philosophy) | Phenomenology. | Content (Psychology)
Classification: LCC BD418.3 .H88 2017 | DDC 128/.2--dc23 LC record available at https://lccn.loc.gov/2016039862

10 9 8 7 6 5 4 3 2 1

For the children of the revolution,
Natura non facit saltum

Contents

Preface   xi
Acknowledgments   xxiii
Abbreviations   xxvii

I

1   Revolution in Mind?   1
    E Is the Word   1
    Old School Cognitivism   3
    Degrees of Radicality   4
    With and without Content   10
    Naturalist Rules of Engagement   13

2   Reasons to REConceive   21
    Equal Partners   21
    Continuity and Break   26
    Less Can Be More   32
    A Radical REConceiving   35
    Handling the Hard Problem   41

3   From Revolution to Evolution   55
    REC's Positive Program   55
    A Certain Take on Predictive Processing   57
    Bootstrap Heaven or Hell?   67

4 RECtifying and REConnecting   75
  RECtifying   75
  Making Sense of Sense Making   75
  Keeping Affordances Affordable   82
  REConnecting   88

5 Ur-Intentionality: What's It All About?   93
  Getting to the Bottom of Intentionality   93
  Ur-Intentionality: The Natural Explanation   104
  Objects and Objections   114

6 Continuity: Kinks Not Breaks   121
  Getting Radical about the Origins of Content   121
  REC's Fatal Dilemma?   122
  Evolutionary Discontinuity?   128
  Kinky Cognition: A Sketch of a Possible Story   137

II

7 Perceiving   147
  Out of the Armchair   147
  Once More unto the Predictive Breach   150
  Integration and Interface   163
  Basic Perceiving Meets Content   171

8 Imagining   177
  Beyond REC's Reach?   177
  Trouble in Mind! Imagine That   183
  A Hybrid, Pluralist Solution: Two Takes   188
  Basic Imaginings at Work: When REC Met MET   193

9 Remembering   203
  Memory's Many Kinds   203
  Enactive, Embodied RECollections   204

Narrative Practice and Autobiographical Memory 206
The Puzzle of Pure Episodic Remembering 215
Roles and Functions of Remembering 221

Epilogue: Missing Information? 233
Don't Mess with Mr. In-Between! 233
Neurodynamics 236
Extensive Dynamics 245
Loops into Culture 253

Notes 255
References 283
Index 315

Preface

> Our entire picture of the world has to be altered even though the mass changes only by a little bit. This is a very peculiar thing about the philosophy, or the ideas, behind the laws. Even a very small effect sometimes requires profound changes in our ideas.
>
> —Richard Feynman, *The Feynman Lectures on Physics*

Classical physics got it wrong. Mass isn't a constant, independent of speed—it increases with velocity, but only appreciably as its velocity approaches the speed of light. Does this matter? Feynman's answer is clear: "Well, yes and no. For ordinary speeds we can certainly forget it and use the simple constant mass law as a good approximation. But for high speeds we are wrong, and the higher the speed, the more wrong we are. Finally, and most interesting, philosophically we are completely wrong with the approximate law" (Feynman, Leighton, and Sands 1963, p. 1–2).

Of course, discovering we are completely wrong philosophically is what spurs on truly "profound changes in our ideas." Feynman concludes that sometimes revolutions in thought ensue from what may seem, for most practical purposes, only small or marginal changes to a theoretical framework.

We couldn't agree more. In the spirit of moving philosophy and science ahead by making well-targeted adjustments to familiar ways of conceiving of mind and cognition, this book starts where its forerunner, *Radicalizing Enactivism: Basic Minds without Content*, left off. Our previous effort was devoted to promoting the fortunes of a Radically Enactive, Embodied account of Cognition, aka REC—an account that conjectures that there could be, and very probably are, forms of cognition without content.

REC holds that some forms of cognition are content-involving in the sense that they represent the world in ways that might not obtain—that is, they represent it in ways that can be true or false, accurate or inaccurate, and so on. Yet it denies that the most fundamental forms of cognition involve contentfully representing the world or being contentfully informed about it in the sense of instantiating correctness conditions of some kind.

Making this twist to how we think about cognition is, from some angles, only a small adjustment, but it is also one that—as we aim to demonstrate in the pages ahead—if accepted, could profoundly change our thinking about thinking.

In distinguishing basic, contentless from content-involving minds, REC seeks to tell the story of mind in duplex terms—as a multi-storey story. In this REC opposes, and is flanked by, more common single-storey stories. It is flanked on the right by accounts of cognition that hold that Cognition always and everywhere Involves Content, aka unrestricted CIC. And it is flanked on the left by those theories that seek to eliminate content across the board—for example, Really Radical Enactive, Embodied accounts of cognition, or RREC.

The changes REC aims to install in the way we think about thinking require theoretical adjustments to our conception of cognition, not mere verbal tweaks. Notably, some philosophers—such as Huw Price—substantially agree with REC in thinking that there are two types of representation at large in cognition. Thus Price recognizes that there is a fundamental difference between responding to and keeping track of covariant information and making contentful claims and judgments that can be correct or incorrect. As he makes clear,

> These two notions of representation should properly be kept apart, not clumsily pushed together. It takes some effort to see that the two notions of representation might float free of one another, but I think it is an effort worth making. ... Once the distinction between these two notions of representation is on the table, it is open to us to regard the two notions as having different applications, for various theoretical purposes. (Price 2013, 37)

Although Price and REC stand terminologically apart, we are together theoretically. It is unimportant that Price uses *representation* as a common label to describe what lies at the heart of two essentially different kinds of cognitive activity. Channeling our inner Freges, we think using the same label in such cases may invite confusion within the sciences of the mind. Still, in the end we follow Shakespeare on this score: Roses by other names!

Once its implications are recognized it is easy to see how REC's proposal excites fundamental disagreements about the character of cognition and the substantive properties that different forms of it are thought to have. For example, REC firmly disagrees with the unrestricted CIC view, now extremely popular in some quarters, that brains, oculomotor systems, and scientists, in doing their primary work, are all doing the same thing—that is, they all put forward contentful hypotheses about how things stand with the world (see, e.g., Hohwy 2013; Gerrans 2014; Clark 2016).

By REC's lights, this is a mistaken view. According to REC, the basic sorts of cognition that our brains help to make possible are fundamentally interactive, dynamic, and relational. REC's signature view is that such basic forms of cognition do not involve the picking up and processing of information that is used, reused, stored, and represented in the brain. The usual form of what REC calls basic, contentless cognition is nothing short of organisms actively engaging with selective aspects of their environment in informationally sensitive, spatiotemporally extended ways. The complex and cascading neural activity that enables this engagement does not involve representing how things stand with the world, but only anticipating, influencing, and coordinating responses in a strong, silent manner.

In promoting its peculiar bifold vision of cognition, *Radicalizing Enactivism* advanced a series of arguments explicitly targeting opposing views, giving the lion's share of attention to the thesis of unrestricted CIC. That book placed a high priority on critiquing mainstream cognitive science's foundational commitment to content-based information processing accounts of mind. It did so because raising doubts about the truth of unrestricted CIC was necessary in order to make the logical space for, and to motivate taking seriously, REC's positive vision of cognition.

Consequently, it is easy to see why those working in the field, such as Sutton (2015), have criticized REC for having an unhealthy concern with providing negative conceptual critiques. Speaking of REC, in his estimation, "In the context of enactivist philosophy ... the engagements with science are too heavily weighted toward the critical mode ... [where] cognitive science is discussed primarily to correct its conceptual confusions" (p. 412).

REC's putative obsession in this regard, Sutton holds, is the reason it has failed, to date, to contribute positively to the many progressive developments in the sciences of the mind. As he sees it, by targeting the question of representational content to the exclusion of all else, RECers have systematically missed out on opportunities to positively connect with and contribute to forward-looking developments in cognitive science.

We are beyond that now. Our radicalizing manifesto was always conceived of as a prolegomenon, one that prepared the ground for a more positive account. Its arguments opened the door to positively rethinking cognition: now, we intend to step through it.

This book decidedly accentuates the positive. Still, as Johnny Mercer's famous song reminds us, accentuating the positive (promoting REC's fortunes) is just the flipside of eliminating the negative (exposing the problems with unrestricted CIC). And, of course, we don't want to mess with Mister In-Between (embracing a Conservative Enactive, Embodied account of Cognition, or CEC). Hybrid accounts tend to inherit weaknesses rather than resolving fundamental problems—so, where possible, it is best to steer clear of CEC's halfway-house proposals.

What is required in order to get our understanding of cognition on a positive footing? Presumably, the optimal account would do all the required explanatory work without adding any superfluous—and potentially distracting—ornaments.

But how does one know what to keep and what to remove? Such decisions can be tricky. Compare the choices that have to be made in designing an automobile for entry in a highly competitive race. Assessing whether to add certain features to the car—such as spoilers—to improve its chances of being first across the finish line is a complicated business, one that requires a balanced assessment of all relevant factors.

There are always pros and cons to take stock of when deciding what to keep and what to remove in order to maximize results. Designers aiming for superior automobile aerodynamics generally seek to reduce the coefficient of drag, a vehicle's overall resistance to airflow. Spoilers generate negative lift—a downforce—and thus increase drag, but in doing so they can improve grip on the road. Knowing whether such additions are a boon or bane in any given case cannot be decided without getting into the devilish details.

This should remind us that there is no a priori reason to suppose, in keeping with the principle of Occam's razor, that slicing away elements from a theory is necessarily a negative move: slicing away does not always result in weaker or less positive explanatory products. Sometimes less is more. Sometimes the leaner car wins the race.

Deciding what a satisfactory account of cognition should include is no less delicate a business than deciding what the most efficient automotive design of a race car should be. Both require making careful, painstaking, all-things-considered assessments.

Victors write histories. But no one is in a position to write the history of cognitive science yet. So we must be careful not to prejudge outcomes. And, in this regard, it is important to be mindful that evaluating the worth of a theoretical framework is not an all-or-nothing affair. Even a framework that gets things wrong in fundamental respects can be of practical and scientific value. Such a framework might provide productive and useful insight into some phenomena, even if this insight proves limited—even if such a framework only allows us to see what we are dealing with through a glass darkly.

There are explanations, and then there is the best explanation. The latter is, of course, the gold standard. How to get at the best explanation of cognition? Philosophy has a major part to play in that enterprise. Philosophy and science must be productive partners if we are to fashion a view of cognition that is without explanatory gaps—one that is empirically adequate and conceptually elegant.

Philosophical work is needed to move our understanding of cognition forward, but deep-seated philosophical convictions can also hinder progress. They can distract and detract. We must beware of unchecked and unsupported philosophical assumptions. As we reveal many times in this book, we must be especially on guard against a priori intuitions that are products of "musty" thinking—intuitions about what cognition "must" be like. This is especially so when these intuitions masquerade as legitimate naturalistic demands on theorizing.

Confusion on this score is the greatest single obstacle to providing sophisticated analyses and open-minded investigations into how best to understand cognition, looking at it both high and low. Such work, done properly, requires adopting—as best we can—a perspective from which all things are considered, and attending to, rather than shying away from, matters of deep theory.

The six chapters of part I aim to clarify REC's duplex account of cognition and to motivate its acceptance.

Chapter 1 sets the scene. It provides a sliding-scale analysis of the degree to which E-positions deviate from the traditional assumptions of cognitivism, revealing why and in what sense REC is radical. It also sets out the basic rules of naturalistic play, reminding the reader why attempts to dismiss REC by appeal to a priori intuitions about what is essential to cognition violate the methodological scruples of naturalism.

Chapter 2 introduces REC's Equal Partner Principle, according to which invoking neural, bodily, and environmental factors all make equally important contributions when it comes to explaining cognitive activity. In line with that principle, it is made clear how REC can accept that cognitive capacities depend on structural changes that occur inside organisms and their brains, without understanding such changes in information processing and representationalist terms.

This chapter also sees the return of the Hard Problem of Content, aka the HPC, which made its debut in *Radicalizing Enactivism*. The HPC is an intractable theoretical puzzle for those explanatory naturalists who hold that information can be distilled from the world through environmental interactions, where such distillation contentfully informs concrete representational vehicles. It is revealed how the need to deal with the HPC can be avoided by adopting REC's revolutionary take on basic cognition, and why going this way has advantages over other possible ways of handling the HPC.

Chapter 3 explicates REC's modus operandi of attempting to incorporate the best resources from other existing accounts of cognition, representationalist and antirepresentationalist alike, to augment its positive explanatory framework. This incorporation is made possible by RECtification—a process through which the target accounts of cognition are radicalized by analysis and argument, rendering them compatible with REC. The RECtification of the Predictive Processing account of Cognition, or PPC, is offered as a shining example of how this procedure works in action. We show how the central ideas of PPC can be given a

REC rendering by abandoning standard cognitivist interpretations, and why this crucial adjustment to PPC is theoretically well-motivated and justified.

Chapter 4 provides further examples of RECtification, this time with the aim of showing how REC can fruitfully ally with and strengthen two prominent nonrepresentational E-approaches to cognition—Autopoietic-Adaptive Enactivism and Ecological Dynamics. These examples of RECtification reveal REC's capacity to marshal and combine powerful resources for explaining basic minds in naturalistic terms. The chapter concludes by discussing the need to show how basic, contentless minds can connect with contentful minds. Doing so is necessary in light of REC's commitment to two ideas: that some cognition is content-involving and that organisms become capable of content-involving cognition by mastering special sociocultural practices.

Chapter 5 explains how it is possible to make sense of REC's proposal that basic minds are contentless while nonetheless holding on to the claim that such minds exhibit a kind of basic intentionality. It situates REC's notion of Ur-intentionality within the larger history of attempts to explicate the notion of intentionality simpliciter, showing that there is conceptual space for a nonrepresentational understanding of intentionality.

The second part of the chapter provides a fresh analysis of how and why this most basic kind of intentionality can be best accounted for in naturalistic terms by means of a RECtified teleosemantics—one stripped of problematic semantic ambitions and put to new and different theoretical use, namely, that of explicating the most basic, nonsemantic forms of world-involving cognition.

Chapter 6 sets out REC's core commitments concerning content-involving cognition and lays out the broad outlines of its proposed explanation for the Natural Origins of Content, or NOC. The chapter also defuses critics' concerns about REC's NOC program in order to establish that it is a tenable way of explaining the natural emergence of content and where content can be found in nature. This requires showing that REC's NOC proposal is neither defeated by the HPC, nor entails evolutionary discontinuity. The chapter concludes by giving a basic sketch of how the NOC program might be pursued, paving the way for further research.

Having cleared the theoretical air, the four chapters in part II pick up the gauntlet thrown down by REC's skeptics and critics and show the explanatory advantages of adopting its duplex vision of cognition. REC puts its positive story into action, showing how its unique resources provide powerful means for understanding perceiving, imagining, and remembering without introducing any scientific mysteries into the mix.

Chapter 7 opens with a reminder that the only properly naturalistic way of debating about the nature of cognition is to stay firmly focused on what is required to explain the relevant phenomena. The chapter then looks again at PPC, in a bid to refute claims that, as matter of fact, tenable versions of PPC need to make indispensable appeal to mental representations. It attempts to defuse arguments that the explanatory punch of PPC requires characterizing perceptual processes and products in representational terms. Such work is necessary, for if that should prove true then REC's attempted appropriation of the main explanatory apparatus of PPC would be thwarted. The final part of the chapter shows how REC, when understood properly, can adequately explain how intramodal and intermodal forms of perceiving can interface and integrate with content-involving modes of perceiving without representational contents forming part of the basis for such explanations.

Chapter 8 begins by arguing that there is no naturalistically respectable way to rule out the possibility of contentless imaginings on purely analytic or conceptual grounds. Hence there is no a priori barrier to understanding contentless, purely sensory-based imaginations in terms of perceptual reenactments involving simulations that are wholly interactive and nonrepresentational in character. Indeed, it is argued that when it comes to understanding basic sensory imaginings and how they manage to do their cognitive work, the focus needs to be on how such imaginings acquire their anticipatory and interactional profiles through embodied engagements with worldly offerings.

The chapter also defends the view that the specific content and correctness conditions of nonbasic, hybrid imaginative attitudes only arise from a combination of basic, purely contentless sensory-based imaginings and the surrounding contentful attitudes of imaginers. Such hybrid states of mind have the right properties to explain the many and varied kinds of cognitive work that imaginings do for us in our daily lives.

Chapter 9 adopts a similar strategy and explores how REC's duplex account of cognition yields special advantages for understanding the many and varied forms of memory—enactive, embodied procedural remembering; pure episodic remembering; and narratively based autobiographical remembering. The chapter argues, on empirical and theoretical grounds, that autobiographical memory is not only content-involving but is a perfect example of a kind of cognition that depends on the mastery and exercise of narrative capacities. In defending this strong claim about autobiographical memory, drawing on REC's understanding of contentless imaginings, it is shown how it is possible to make sense of pure episodic forms of remembering that operate before and below the capacity to autobiographically narrate the past. The chapter concludes by considering general arguments, motivated by empirical findings, that compel a rethinking of purely CIC representationalist and content-based views of the primary function of remembering.

The epilogue takes a last look at the possibility that REC may be leaving out something explanatorily important because it says nothing about information that many believe is acquired, processed, pooled together, mapped and remapped, and generally made use of by the brain. It is argued, focusing on a prominent case in point, that even the groundbreaking research on the positioning systems in rat brains can be accommodated within the REC framework—which assumes the brain is informationally sensitive but does not process informational content—without explanatory loss and with the explanatory gain of not having to deal with the HPC. It is then shown how the view about neurodynamics that REC recommends is wholly compatible with REC's position that cognitive phenomena are fundamentally extensive and world-involving, such that it is possible for some minds, at least, to loop into society and culture and vice versa.

In all, our efforts in the chapters of this book illustrate the positive advantages of adopting REC's duplex vision of mind and the many ways in which enactivism "keeps evolving by incorporating new empirical studies and theoretical perspectives" (Colombetti 2014, p. xiv).

Acknowledgments

We are sincerely and humbly grateful to our families—the Huttos, Farah, Alex, Justin, and Emerson and the Myins, Inez, Elise, Charlotte, and Laure—for their support during the writing of this book. We deeply apologize for all the impositions and inconveniences we caused during the more intense phases of the project. As a consequence of Dan's move to Australia in 2013, our face-to-face collaboration required one or the other of us to be away from home for frequent lengthy periods. We could not have brought this book into being without such side-by-side interactions, but our families had to make many sacrifices to enable them. We owe our loved ones a unique debt that simply cannot be repaid.

We are especially grateful to the University of Antwerp and the University of Wollongong, both for the funding they provided to support our joint work and for approving extended leaves of absence that made our international collaborations possible. We thank the Australian Research Council for funding the Discovery Project "Minds in Skilled Performance" (DP170102987) which supported this research and the Research Foundation Flanders (FWO-Vlaanderen) for helping fund the projects Computation Reconsidered (G0B5312N),

Offline Cognition (G048714N), and Getting Real about Words and Numbers (G0C7315N), as well as for sponsoring Erik's three-month research stay in Wollongong (V404715N). We also thank the Research Council of the University of Antwerp for the DOCPRO project Perceiving Affordances that informed our thinking about perception.

For their helpful suggestions, comments, challenges, criticisms, and other contributions along the way, we thank Nik Alksnis, Louise Barrett, Filip Buekens, Tony Chemero, Andy Clark, Giovanna Colombetti, Hanne De Jaegher, Ezequiel Di Paolo, Catarina Dutilh Novaes, Martin Fultot, Shaun Gallagher, Vittorio Gallese, Dui Garofoli, Paul Griffiths, Matthew Harvey, David Kaplan, Fred Keijzer, Michael Kirchhoff, Julian Kiverstein, Richard Menary, Danièle Moyal-Sharrock, Fred Muller, Dominic Murphy, Bence Nanay, Alva Noë, Kevin O'Regan, Ian Ravenscroft, Erik Rietveld, Glenda Satne, Bilge Sayim, Marc Slors, John Sutton, Jasper van den Herik, Tilde Van Uytven, Mike Wheeler, and Rob Withagen.

We thank our departmental and faculty colleagues and PhD students at the Universities of Antwerp and Wollongong for engaging so actively with our work, with a very special thanks to Karim Zahidi for his rapid checking and proofreading of the final manuscript just before its submission, and to Jan Van Eemeren and Farid Zahnoun for help with references and additional proof-reading.

We are grateful to the organizers of and audiences at the many events at which we have presented our individual work worldwide. We would especially like to thank Jingkun Chen, Lucas Thorpe, Tobias Schlicht, Pierre Steiner, and Tomasz Komendzinski for hosting workshops, talks, and extended lecture series—in Taiyuan, China; Istanbul, Turkey; Bochum,

Germany; Compiègne, France; and Torun, Poland, respectively. These events focused exclusively on our evolving thoughts about REC and gave us an opportunity to come together to clarify and develop ideas that have proved crucial to the main themes and arguments of this new book.

Several other conferences and events that we attended individually were especially important in the development of our ideas. We thank the organizers of these events. For Erik, these include New Directions in Philosophical Psychology in Milan, September 2015 (Antonella Corradini); ASSC Workshop on Sensorimotor Theory, Paris, July 2015 (David Silverman and Jan Degenaar); and Embodied Design in Education, Utrecht, October 2015 (Arthur Bakker). For Dan, these include Varieties of Enactivism—A Conceptual Geography, London, April 2014 (David Ward, David Silverman, and Mario Villalobos); Arguing with Dan Hutto, Lisbon, June 2015 (Rob Clowes and Dina Mendonça); and Culture, Extended and Embodied Cognition and Mental Disorders, Helsinki, June-July 2016 (Pii Telakivi, Anna Ovaska, and Tuomas Vesterinen). Indeed, thanks to the generosity of the organizers of the last event on that list I (Dan) sit penning the very final words of this book while waiting for a flight home to Oz in the very comfortable Töölö Towers.

We are beholden to the editors and publishers of the following journals and edited volumes for allowing us to reuse material from the articles and chapters listed below, which we have reworked and woven into this book: D. D. Hutto and E. Myin, "Going Radical," in *The Oxford Handbook of Cognition: Embodied, Embedded, Enactive, Extended*, ed. A. Newen, S. Gallagher, and L. de Bruin (Oxford University Press, in press); E. Myin, and D. D. Hutto, "REC: Just Radical Enough," Studies in Logic, Grammar and Rhetoric 41 (1) (2015): 61–71; D. D. Hutto, "REC:

Revolution Effected by Clarification," *Topoi* (2015): DOI 10.1007/s11245-015-9358-8; D. D. Hutto, "Getting into the Great Guessing Game: Bootstrap Heaven or Hell?," *Synthese*, in press; D. D. Hutto and G. Satne, "Continuity Skepticism in Doubt: A Radically Enactive Take," in *Embodiment, Enaction, and Culture*, ed. C. Durt, T. Fuchs, and C. Tewes (MIT Press, in press); D. D. Hutto and G. Satne, "The Natural Origins of Content," *Philosophia* 43, no. 3 (2015): 521–536; D. D. Hutto, "Overly Enactive Imagination? Radically Re-Imagining Imagining," *Southern Journal of Philosophy* 53, no. S1 (2015): 68–89; D. D. Hutto, "Memory and Narrativity," in S. Bernecker and K. Michaelian, *Handbook of Philosophy of Memory* (Routledge, in press).

We thank Philip Laughlin and the team at the MIT Press for being such good sports and so accommodatingly flexible with us in getting this long-delayed story into print.

Finally, for helping to materially fuel our cognitive efforts, we thank the friendly staff of Elzenveld Centrum and Kapitein Zeppos café: the official homes of REC in Antwerp.

Abbreviations

AAE	Autopoietic-Adaptive Enactivism
CIC	Content Involving account of Cognition
CEC	Conservative Enactive, Embodied account of Cognition
HPC	Hard Problem of Content
MET	Material Engagement Theory
NOC	Natural Origins of Content
PPC	Predictive Processing account of Cognition
REB	Radically Enactive, Embodied account of Behavior
REC	Radical Enactive, Embodied account of Cognition
RREC	Really Radical Enactive, Embodied account of Cognition
SIT	Social Interaction Theory of Autobiographical Memory
Ultra CEC	Ultra Conservative Enactive, Embodied account of Cognition

1 Revolution in Mind?

You say you want a revolution
Well, you know
We all want to change the world
—The Beatles, "Revolution"

E Is the Word

E is the letter, if not the word, in today's sciences of the mind. E-approaches to the mind—those that focus on embodied, enactive, extended, embedded, and ecological aspects of mind—are now a staple feature of the cognitive science landscape.

A strong motivation that has spurred on the E-movement has been the need to develop theories that can overcome well-known problems encountered when attempting to understand and model the fluid and plastic nature of cognition. These limitations are especially conspicuous when trying to explain the intelligence of fast-paced, spontaneous, but skilled performance in terms of classic reasoning processes involving the manipulation of in-the-head, amodal symbols and propositions (Sutton et al. 2011). Even assuming that such forms of reasoning may be tacit, they are still deemed too slow, rigid, and abstract to properly account for the dynamically updated character of real-time intelligent activity. Researchers in the field have turned to E-approaches for a better characterization of the contextualized sensitivity and responsiveness of such intelligence, thinking of it in terms of embodied, enactive know-how about situations that does not involve positing "a clunky set of internalized propositions" (Sutton and McIlwain 2015, 100; see also Dreyfus 2014).

Another major catalyst for E-theorizing has been the need to accommodate new empirical findings that reveal that a great deal of cognition is—in some centrally important respects—connected, and sensitive, to facts of embodiment. Experimental findings of this sort—those that Goldman (2012) makes much of—include the use of circuits associated with motor control functions in higher-level language comprehension tasks (Pulvermuller 2005); the reuse of motor control circuits for memory (Casasanto and Dijkstra 2010); the reuse of circuits that mediate spatial cognition for a variety of higher-order cognitive tasks (e.g., the use of spatial cognition for numerical cognition) (Hubbard et al. 2005; Andres, Seron, and Olivier 2007); mirroring phenomena, including not only motor mirroring but also the mirroring of emotions and sensations (Rizzolatti et al. 1996; Rizzolatti and Sinigaglia 2010; Keysers, Kaas, and Gazzola 2010); and sensitivity to perceiver's own bodily states when estimating properties of the distal environment (Proffitt 2008).

As part of the larger E-turn, many productive scientific research programs are trying to understand the significance of E-factors for the full range of cognitive phenomena, with new proposals about perceiving, imagining, remembering, decision making, reasoning, and language appearing apace (Wilson and Foglia 2016).

Some hold that these developments mark the arrival of a new paradigm for thinking about mind and cognition, one radically different from cognitive science as we know it.[1] Others maintain that accommodating E-factors, while important, requires only very modest tweaking or, at most, some crucial but still limited revisions to the business-as-usual cognitive science framework. By conservative lights, radicals vastly exaggerate the theoretical significance of the so-called E-turn. Moderates hold that whatever changes may be required they will fall short of reconceiving cognition.

Old School Cognitivism

Before assessing the scale and magnitude of the theoretical changes that may need to be wrought in order to properly accommodate E-findings, it is important to be clear about which traditional assumptions are potentially at stake. That task entails getting clear about the central tenets of cognitivism, which has enjoyed the status of the default approach for conceiving of cognition in the sciences of the mind since the 1950s.

Contemporary cognitivism takes it to be axiomatic that "the mind represents and computes" (Branquinho 2001, xv). In doing so it endorses an intellectualist vision of minds that made its debut in early modern times, making representationalism and computationalism the two main pillars of cognitivism.[2]

These twin pillars of cognitivism rest on a more foundational substratum. Cognitivism is methodologically committed to providing explanations of a mechanistic variety. According to the early modern take on mechanistic explanation, explaining a phenomenon always involves two steps. First, it is necessary to identify, through analysis, component parts and their principles of interaction. Second, it is necessary to show, through synthesis, how the interactions between such parts generate some phenomena (Horst 2007, 16–17).

In a similar vein, today's mechanists also emphasize the need to discover the parts, operations, and organization of the mechanisms that underlie and causally generate phenomena of interest (Bechtel 2008; Bechtel and Richardson [1993] 2010; Craver 2007; Kaplan 2015). In the sciences of the mind this general idea is embraced in a specific version, namely the assumption that cognitive mechanisms are distinct from, and produce, intelligent behavior. This assumption "was among the defining features of the cognitive revolution" (Aizawa 2015, 759).

Today's cognitivism tends to make a further, more specific assumption about the location of the mechanisms responsible for intelligent activity—namely that cognitive processes that give rise to such activity take the form of brain-based computations over internal mental contents. A familiar line of thought is that if one allows that "cognition is a cause of behavior, one can better appreciate why it might be something realized in the brain alone" (Aizawa 2015, 756).

To assume that representational-computational mechanisms are neural is to endorse an I-conception of mind that is methodologically and metaphysically committed to Individualism, Intellectualism, and Internalism. From such a perspective, cognition only goes on in the intellectual interior of individuals.

Degrees of Radicality

With these reminders about the core features of the dominant cognitivist framework in place, it is possible to gauge to what extent and precisely how existing E-theories are more or less conservative or revolutionary with respect to it.

Some E-theorists see no need for any revisions to standard cognitivism—neither to the two pillars of cognitivism nor to the I-conception of mind. In attempting to make good on this idea they argue that cognition can be, and often is, grounded in embodied representations—representations whose content is about the body and is carried by vehicles in embodied formats (Goldman and de Vignemont 2009; Gallese and Sinigaglia 2011; Alsmith and de Vignemont 2012; Goldman 2012, 2014; Gallese 2014).

E-theories of this kind offer an Ultra Conservative Embodied account of Cognition, or Ultra CEC. Ultra CECers attempt to accommodate the wealth of empirical findings about the contribution E-factors make to cognition while still holding on to the idea that cognition is wholly representational-cum-computational and grounded in entirely brain-based mechanisms.

They are persuaded by a wealth of evidence showing that, in Goldman's words (2012, 72), "embodiment would seem to be realized to a significant degree, a degree quite unanticipated by cognitive science of two or three decades ago."

By calling on the construct of embodied representations, Ultra CEC theorists claim to have the resources needed to explain how and why embodied cognition is so pervasive. They attempt to do so by appealing to the fact that the brain gets a great deal of its cognitive work done by reusing or redeploying embodied representations for many and varied cognitive tasks.

In essence, the Ultra CEC assumption is that "cognition is embodied in the sense that the mechanisms for perception and action are the same as the mechanisms for concept manipulation and reasoning" (Aizawa 2015, 758). If embodied representations with the aforementioned properties exist, then it is potentially possible to explain how and why cognition sensitive to embodied factors plays such a prevalent role in so many cognitive domains. Through this means, Ultra CECers hold that the real work of cognition is surprisingly often E-ish while still only a matter of the manipulation of representations in the brain. For their approach to work, Ultra CECers must make good on the idea that some neural representations have rather special E-ish contents and formats.

Other E-theorists are more daring in moving away from old school cognitivist commitments. They hold that nonneural, temporally extended embodied engagements can feature in, and perhaps even constitute, cognition. A familiar version of such an approach, promoted by advocates of the extended mind hypothesis, is that the vehicles of cognition might sometimes extend across the nonneural body and environment and bear some of the cognitive load in enabling the completion of specific tasks (Clark and Chalmers 1998; Clark 2008b; Rowlands 2009; Wheeler 2010).

Extended mind theorists heavily stress the transformative potential of external tools, which range from spoken to written words and other symbols, computers, and actual or possible bodily extensions such as brain implants. The production of abstract art is a case in point (as discussed in Clark 2003; see also Myin and Veldeman 2011). Research by van Leeuwen, Verstijnen, and Hekkert (1999) on the role of sketchpads in the production of certain forms of abstract art reveals that the creation of multiply interpretable elements in drawing, typical in such art, is essentially dependent on the use of external sketches.

This is due to limitations of our biologically unsupported imaginative capacities. Producing certain kinds of abstract art requires dealing with multiple and competing interpretations of images simultaneously. But it is apparently not possible to hold such images in the mind at once (Chambers and Reisberg 1985). Without supplement, our natural mental equipment is simply not up to that task by itself. The production of art of the relevant kind depends on external support: making use of sketched images is necessary to guide the design process in a reliable way. The use of sketchpads makes certain forms of artistic creation possible—artforms that would not have existed otherwise.

On the standard extended functionalist interpretation, the environmental contributions that make such artistic production possible are best understood in terms of extended vehicles of cognition—extended vehicles that play specific computational roles as part of larger information processing mechanisms (Clark 1997, 2008a, 2008b).

Importantly, a guiding assumption of extended mind theorists is that in "limiting embodied cognition to some sort of information processing ... it is not that literally any causal contributor to performance realizes cognition ... [but] only causally relevant informational contributions to performance realize cognition" (Aizawa 2015, 769). Proponents of the extended functionalist view of mind thus retain the idea that cognition consists in representationally informed computational processes, but unlike advocates of Ultra CEC, such E-theorists do not assume such processes are always wholly neural and brainbound.

Sensorimotor enactivism, as canonically formulated in Noë 2004, goes further still in breaking faith with traditional cognitivist thinking.[3] It holds that central forms of cognition are constituted by and supervene on wide-reaching, temporally extended, interactive embodied engagements with the world.

Yet it steers clear of any form of computational functionalism. Committed to the idea that idiosyncratic features of our embodiment matter to the character of our cognition, it abandons a central tenet of functionalism—the thesis of multiple realizability—in holding that "to perceive like us ... you must have a body like ours" (Noë 2004, 25).[4]

Whereas extended functionalism allows that internal models sometimes underpin perception, sensorimotor enactivism rejects the idea that we form rich and detailed inner representations when perceiving. Sensorimotor enactivism's big idea is that perceiving "isn't something that happens in us, it is something we do" (Noë 2004, 216; see also O'Regan and Noë 2001). Although it deems activity in neural substrates to be necessary for perceiving or having perceptual experience, perceiving is nevertheless understood to be "realized in the active life of the skilful animal" (Noë 2004, 227; see also Silverman 2013). Even so, as originally formulated sensorimotor enactivism remains conservative in clinging to the idea that "for perceptual sensation to constitute experience—that is for it to have genuine representational content—the perceiver must possess and make use of sensorimotor knowledge" (Noë 2004, 17).

Despite their substantial challenges to the tradition, both the extended mind thesis, construed under the auspices of functionalism, and sensorimotor enactivism are Conservative Enactive, Embodied accounts of Cognition, or CEC. Extended functionalism is the more conservative of the two positions when it rests on the two pillars of cognitivism, even though it relinquishes a complete commitment to the I-conception of mind. Sensorimotor enactivism goes much further, not only in giving up on the I-conception more thoroughly than extended mind theorists, but also in questioning computational functionalism. Still, it too—at least in Noë's (2004) rendering—maintains representationalism about the character of even the most basic kinds of cognition, such as perception.

The most radical view—the one that makes the cleanest break with the cognitivist tradition—is a Radically Enactive, Embodied account of Cognition, or REC. REC theories ask us to rethink—root and branch—old school conceptions of cognition, demanding that we revise our views of the mind's core work and how it gets done. Like sensorimotor enactivism, REC theories understand cognition as something that organisms do. Cognition is a kind of embodied activity that is out in the open, not a behind-the-scenes driver of what would otherwise be mere movement. REC theories conceive of the basis of cognition in terms of extensive and dynamically loopy processes that are responsive to information in the form of environmental variables spanning multiple temporal and spatial scales.

Crucially, REC theories construe cognition as unfolding, world-relating processes rather than as a series of content-bearing states and their interactions. Cognitive processes unlike states have spatial reach and unfold over time. Importantly, unlike a state or event, a process is "something which goes on through time and can change as it does so" (Steward 2016, 76).[5]

Our version of REC is part of a wider movement, one inspired by scientific developments in robotics, dynamical systems theory and ecological psychology and which finds philosophical support from the phenomenological, American naturalist, and Buddhist traditions of thought. In recent years this movement established itself most prominently through the seminal work of Varela, Thompson and Rosch (1991). REC is thus part of a larger family of approaches to cognition encompassing any and all which maintain that (1) cognition is a kind of situated enactive, embodied activity, and that (2) enactive, embodied activity does not always and everywhere involves thinking about the world in contentful ways.

Embracing these two tenets is the minimal commitment—the lowest common denominator—of any REC-style view. These twin tenets are the theoretical core—the philosophical nucleus—shared by a wider set of radically embodied, enactive and ecological theories, each of which is distinguished by the specific explanatory resources that they bring to the table (Thompson 2007; Di Paolo 2009; Chemero 2009; Froese and Di Paolo 2011, Hutto and Myin 2013, Bruineberg and Rietveld 2014).

With and without Content

Supporters of REC disagree with CECers on a pivotal issue about the nature of basic minds. RECers deny that all forms of cognition, and in particular its root forms, are content-involving. It rejects the Content Involving account of Cognition, or CIC, in its unrestricted form.

What is the central notion of content that radically minded RECers deny is a feature of basic minds? It is any notion of content that assumes the existence of some kind of specified correctness condition. To be in a contentful state of mind is to take ("represent," "claim," "say," "assert") things to be a certain way such that they might not be so. This generic idea of content is the notion that analytic philosophers and classical cognitive scientists commit to when they suppose that cognizers "represent things as being thus and so—where, for all that, things need not be that way" (Travis 2004, 58).

It is usual for analytic philosophers of mind to assume that content so understood equates with propositional content.

Brogaard (2014), for example, tells us that "perceptual experience is accurate or inaccurate. If it's accurate, it's accurate in virtue of some proposition p being true. If it's inaccurate, it's inaccurate in virtue of some proposition p being false. But that *proposition p just is the content* of perceptual experience" (p. 2, emphasis added).

Brogaard's comments give voice to the pervasive tendency among analytic philosophers to understand content in essentially propositional terms. Thus in the passage above she equates accuracy conditions with truth conditions. Yet the notion of content is elastic enough to allow that the relevant correctness conditions might be understood in terms other than truth: say, in terms of accuracy, veridicality, or some other kind of satisfaction condition where these are taken to differ from truth conditions (see, e.g., Crane 2009; Burge 2010).

The notion of content that REC denies is a feature of basic minds is therefore somewhat elastic—but it is not so elastic as to include every conception of content that abounds in the philosophical literature.

For example, it does not automatically include what is sometimes called phenomenal content. It cannot be taken for granted that to enjoy an experience with a certain phenomenal character is to be in a state of mind with representational content. A great deal of argument would be needed in order to establish such a reduction or identity.

Nor, in a similar vein, can it be simply assumed that contentful states of mind are always in play whenever an agent stands in a cognitive relation to, or has attitudes directed at, specific objects or states of affairs. Here too, some philosophers—especially those inspired by the phenomenological tradition—speak of intentional content when describing such states of mind. But it does not follow that the notion of intentional content they invoke reduces to representational content possessing any kind of correctness conditions. For instance, in speaking of intentional content Dreyfus (2002, 414) draws a deliberate contrast between states of mind that are merely world-involving and those that possess satisfaction conditions, thus maintaining that "there are inner states of the active body that have intentional content but are not representational."

The foregoing observations reveal that some philosophers use the word *content* so liberally that it just picks out the object of experience, perception, or thought, whereas others use the notion in a restrictive sense that entails the existence of some kind of satisfaction conditions. These two uses must not be conflated. Certainly, lax and liberal use of the notion of content should not mask the fact that a great deal of argument would be needed to establish that all acts of world-engaging experience, perceiving, or thinking involve contents with conditions of satisfaction.

REC assumes that some cognitive attitudes are contentful in the restrictive sense of possessing correctness conditions. REC holds, for example, that we sometimes think thoughts that refer to things beyond themselves—thoughts that can be true or false. Nevertheless, it denies that having thoughts with content—so understood—is fundamental to all cognition. By REC's lights, acquiring the capacity for cognition that involves content is a special achievement. Creatures capable of contentful cognition, in the REC view, will have had to master very special kinds of scaffolded practices—practices involving public norms for the use of symbols, where such norms depend for their existence on a range of customs and institutions (see Hutto and Satne 2015).

REC thereby distinguishes basic from nonbasic minds by appeal to the requirement that the latter are content-involving. It takes the former, elementary kinds of mind to be phylogenetically and ontogenetically fundamental. Importantly, REC holds it is possible to go quite a long way, cognitively speaking, without involving content in the specified sense. For example, not all kinds of culturally shaped acts of cognition are content-involving. Being influenced in what one perceives to be a threat and the way one does so can be culturally shaped without always and necessarily involving content.

Crucially, REC holds that there are interesting varieties of basic perceiving, imagining, and remembering—which can come in the form of embodied activity or reenactments—that involve no content (Hutto 2014, 2015b, in press). Hence, importantly, being basic minded in the REC sense does not entail that one is operating with only a low-grade form of cognition.

Naturalist Rules of Engagement

Is it coherent to hold that there could be minds that lack content? Is a nonrepresentational theory of cognition possible? Is the notion even conceptually coherent? REC requires conceiving of basic cognition in noncontentful terms. Some believe that is not simply a bad idea but an impossible one: accordingly, the revolution in thinking that REC seeks to bring about isn't just a nonstarter or something to be dismissed on empirical grounds, it is literally unimaginable and thus can be ruled out a priori as inherently conceptually confused.

This will be the reaction of those who take it that bearing mental content is the true mark of the cognitive. Anyone who accepts this condition must deny that the sciences of the mind could ever go radical. There is an apparently serious problem for REC if it is assumed that the existence of mental states that bear content defines and demarcates the subject matter of cognitive science (Shapiro 2014a, 2014b). Those persuaded by this criterion will find it simply unthinkable that cognitive science could ever abandon the idea that basic states of mind are content-involving.

This unrestricted CIC assumption about the mark of the cognitive is widespread and easy to find in the literature. The following quotations capture its core commitments and highlight its perceived importance:

Admittedly, delimiting the scope of the "cognitive" is not an easy matter, but ... it seems adequate to specify that cognitive states, structures, and capacities are mental entities with representational content. (Khalidi 2007 93)

Without representation cognitive science is utterly bereft of tools for explaining natural intelligence. We would go further: without representation there is no cognitive (as distinct from behavioral, biological, or just plain physical) science in the first place. (O'Brien and Opie 2009, 54)

Drawing on this unrestricted CIC assumption about the nature of cognition, it has been claimed that REC cannot possibly provide an account of basic cognition but only, at most, an account of contentless forms of complex behavior. Accordingly, going radical about basic cognition is simply not a live option and any proposed radical rethinking of cognition can be known, in advance, to be not false but incoherent. By this logic, E-accounts are only candidate theories of cognition if they adhere to CEC. As Shapiro (2014a, 219) states, because REC violates this condition it is not clear how it "could be a science of the mind rather than, say, behaviour."

Thus, if the REC revolution were to succeed, cognitive science would have to trade in its sole concern with its traditional subject matter—cognition—and take up an interest in behavior as well. Surely, as Aizawa (2014, 19) says, "That would be a real revolution."

By these lights REC is conceptually debarred from being a genuine rival to an unrestricted CIC account of cognition. If cognition is defined as always and everywhere involving content then, as a matter of logical necessity, the two frameworks must be interested in different explananda—different explanatory targets—and employ different explanantia—different kinds of explanations. If this were so then REC's talk of basic cognition would be a misnomer. By this reasoning, it would be simply impossible for REC to qualify as an account of basic cognition. As such REC proposals about basic cognition would necessarily reduce to REB—a Radically Enactive, Embodied account of Behavior.

Anyone who holds that conceiving of cognition in the absence of content is simply impossible treats unrestricted CIC as an already-known, conceptually based analytic truth.

Is ruling out the very possibility of REC in this analytic, a priori manner justified? Not for naturalists. An analytic defense of unrestricted CIC is not open to anyone who adopts the kind of naturalistic approach to philosophy that cognitive science demands. Attempts to defeat REC by appeal to an unrestricted CIC mark of the cognitive—namely by appeal to that assumption as an axiomatic first principle—violate naturalism by committing a serious methodological foul.

Defending unrestricted CIC by invoking unrestricted CIC as a demarcation criterion that articulates the "mark of the cognitive"—one that defines the subject matter of cognitive science—is blatantly circular. Whether such a move is viciously or virtuously circular is beside the point for, as Ramsey (2014, 4) observes, it is, in any case, "not supported by a proper scientific outlook." Worse still, as Ramsey emphasizes, this move is bound to lead to bad consequences, such as (1) unnecessarily restricting our theorizing about cognition, (2) undermining the empirical nature of the representational theory of mind, and (3) encouraging substantial weakening of the notion of representation.

To illustrate how this demarcation gambit leads to bad outcomes, consider how Noë's evolving views on perception would have to be handled by anyone endorsing the unrestricted CIC demarcation criterion.

Building on O'Regan and Noë 2001, Noë 2004 offered a sensorimotor theory of perceiving, one that centrally incorporated many enactivist insights. As noted earlier, Noë's 2004 theory is clearly CEC. This is because it conservatively retains commitment to the idea that perceiving is content-involving in a fully representationalist sense.[6] Importantly, those who claim that REC must reduce to REB do not see Noë's CEC approach as falling afoul of a similar fate. Why not? Any E-theory of a CEC kind, such theorists hold, is safe to the extent that it endorses unrestricted CIC.[7] Thus Noë's 2004 theory—which holds that perceiving is contentful—qualifies as a bona fide theory of cognition for anyone who plays the unrestricted CIC demarcation card.

There is a problem, though. In subsequent writings, Noë (2009, 2012) has apparently abandoned his earlier CIC take on perception. For example, Noë (2009, 99) advances the view that "to perceive something is not to consume it, just as it isn't a matter of constructing, within our brains or minds, a model or picture or representation of the world without. There is no need. The world is right there and it suffices."

Noë (2012) too shows some clear signs of endorsing REC over CEC (or for the sake of argument, let's just assume that this is so). The question is: Would such a change in thinking entail that Noë's new theory of perception no longer concerns cognition? Does it automatically thereby convert into a theory of behavior?

Surely not. Perceiving is a paradigmatic cognitive phenomenon: to think it becomes non-cognitive and converts into mere behavior when understood by REC's lights as lacking content is absurd. By a similar token, shifting the nature of the explanans from CEC to REC footing doesn't change the target explanandum—what a theory is about and seeks to explain. Even if by going radical the theory should turn out to be empirically false, what would be on offer would still be a failed theory of perception and hence cognition. The only change would be that non-CIC tools would be offered for understanding the same cognitive phenomenon that is of interest to REC's rivals. The moral is that the extensional target of theorizing—what we are interested in understanding as opposed to how we understand it—is not determined by the nature of proposed explanantia.

Another analytic demarcation move is to try to secure the truth of unrestricted CIC in advance of empirical developments by designating that whatever is actually discovered to play the relevant role in explaining intelligent activity must be, by definition, a "contentful representation." Used in this way the label "contentful representation" is guaranteed to pick out anything that, in the end, best explains cognition.[8] The problem with trying to secure unrestricted CIC's truth in this way is that it results in "an utterly vacuous outlook" (Ramsey 2014, 10). Unless the properties content has and the precise role such properties play in explaining behavior are specified, representational theories of mind are rendered empirically empty. Under such conditions, unrestricted CIC theorizing about the mind reduces to a mere wait-and-see game of bestowing the label "representation" on the properties that actually turn out to characterize cognition.

In sum, these considerations illustrate how in various ways adopting an a priori, analytic demarcation stratagem leads to all of the bad consequences identified by Ramsey (2014). Ultimately, such analytic moves break faith with a properly naturalistic methodology. Naturalist theory building is meant to be substantive and speculative: it takes risks and goes beyond pure forms of conceptual analysis. The test of the tenability of a proposal or hypothesis about the nature of cognition is that it accommodates existing data better than rivals. This requires making comparisons with competing theories in order to assess a theory's empirical adequacy and global fit with surrounding theories so as to generate hypotheses and test which theory provides the best explanation (see, e.g., Carruthers 2011, xiii; Sterelny 2012, xi).

The bottom line, as concerns basic cognition, is that a good naturalist cannot both demonstrate the superiority of unrestricted CIC proposals empirically and, at the same time, rule REC out analytically. They must abide by Ramsey's Rule, which legislates that in this naturalistic contest, "You can't treat representational posits as both interesting explanatory constructs *and* as a necessary condition for a legitimate account of the phenomena you are trying to explain" (Ramsey 2014, 8).

It should now be clear that if REC could be ruled out from the philosophical armchair in this way then any such perceived victory for unrestricted CIC traditionalists would be scientifically hollow. To defeat REC in such a dismissive manner would be to sacrifice a win that demonstrates unrestricted CIC's substantive, superior explanatory power for a win by analytic stipulation. The only naturalistically respectable way to defeat REC is to give it its day in empirical court, determining, in the end, whether it or unrestricted CIC offers the best account of various cognitive phenomena, all things considered. This requires active investigations conducted in an open-minded way. We will come to that.

2 Reasons to REConceive

You tell me that it's evolution
Well, you know
We all want to change the world
—The Beatles, "Revolution"

Equal Partners

When it comes to explaining intelligent activity REC subscribes to the Equal Partner Principle. The Equal Partner Principle denies that neural factors have any special cognitive status in such explanations. Citing neural factors carries no greater explanatory weight, for example, than citing bodily and environmental factors. REC advocates the even-handed treatment of all such factors because it conceives of cognitive systems in dynamical terms. From the dynamical perspective, variables of any kind make an equally important contribution, irrespective of where they lie with respect to the boundaries of skin and skull, just as long as they make an appropriate contribution to explaining the overall shape of the system's responsiveness.

On this issue, REC disagrees with ultra conservative views of cognition—those which hold that cognitive explanations are only concerned with what goes on in heads, thereby indirectly conferring a special status on citing neural factors in such explanations. In a more subtle way, REC also disagrees with extended functionalists. This may not be immediately obvious since extended functionalists accept that non-neural factors can have an explanatory status that is on a par with neural factors. Yet extended functionalists hold that non-neural factors enjoy this explanatory parity only to the extent that such factors play special information processing or computational roles. Since REC does not insist on that special condition, its version of the Equal Partner Principle is different from and much more liberal than that proposed by extended functionalists.

An example of a RECish explanation that honors the Equal Partner Principle is found in the constraints-led approach to skill acquisition which is inspired by developments in ecological dynamics.

Ecological dynamics seeks to combine basic principles of ecological psychology and dynamical systems theory. On the one hand, it draws heavily on ideas central to Gibson's (1979) ecological psychology, assuming that there is a tight fit between animals and their environment, and that perception is fundamentally bound up with and is for action. For Gibsonians, perceiving is an active, dynamic process in the service of getting an effective, practical grip on the world. Accordingly, perception takes the form of targeted interactions with aspects of the world, and thus it is extended in both time and space. Notoriously, Gibsonians are adamant that perceiving, so conceived, can be fully explained without the need to posit any intervening inferences or mental representations.

The constraints-led approach augments these core ideas from ecological psychology by calling on unique explanatory resources from dynamical systems theory (Davids, Button, and Bennett 2008; Chow et al. 2011, 2015). Dynamical systems theory is the perfect partner for the Gibsonian conception of perceiving as an embodied activity. This is because it employs differential equations to explain and predict how the states of nonlinear systems evolve over time. It begins by taking stock of a number of variables that describe the state of a system at a particular point in time. It then makes use of its special mathematical tools to chart the trajectory of changes in the states of such systems as they move through a space of possibilities, which is known as a phase space. In the language of dynamical systems theory, nonlinear systems move toward regions that attract them and away from regions that repel them, traversing a landscape known as a topology.

The equations of dynamic systems theory describe the tendencies of such complex systems, including their tendencies for interacting with other such systems. What makes complex, nonlinear systems special is that they are self-organizing—they exhibit an order that is produced and constrained by mutual and reciprocal influence of their components. Yet, crucially, the factors that shape the trajectory of a system do not do so "in the sense of 'instructing' how it should behave or of 'monitoring' its evolution" (Colombetti 2014, 36). Moreover, the effects of such influence are not reliably proportional: "a small (large) change in some variable or family of variables will not necessarily result in a small (large) change in the system" (Rickles et al. 2007, 934).

In these key respects, dynamical systems are importantly unlike linear systems. Linear systems can be functionally decomposed and analysed in terms of their structural parts and respective operations. Functional decomposition of a linear system into the sum of its parts makes it possible to analyse each part separately and thereby to explain and predict how the whole system taken as the sum of those parts will behave over time. Since nonlinear, dynamical systems are not amenable to such decomposition, their evolution over time requires the special mathematical techniques described above.

Combining central ideas from ecological psychology with the special mathematical tools of dynamical systems theory, as in Chemero 2009, has opened the way for fruitfully investigating the complex self-organizing responsiveness of learners acquiring skills in embodied activities. From a constraints-led perspective the processes involved in the mastery of embodied skills can be characterized by a number of interacting variables, and the continuous, temporal, and interdependent changes in their unfolding patterns can be captured by a set of differential equations. Importantly, although it is possible to focus on the dynamics of the parts of complex systems for various purposes, the basic unit of analysis for ecological dynamics is the nonlinearly coupled organism-environment system.

Crucially, from the vantage point of a constraints-led approach, individuals are understood as situated dynamical systems that are simultaneously open to influence and intervention on multiple scales. Training is focused on shifting behavior of such dynamical systems, which are prone to reconfigure, enabling them to self-organize quickly and flexibly to meet contextual demands (Kelso 1995). Therefore, trainers working within this paradigm selectively modify specific bodily, environmental, and task constraints—for example, changing the size of playing fields, adjusting distances between players, causing players to become fatigued—to shape and control the emergence of skills and expertise over time (Newell 1986; Hutto and Sánchez-García 2015).

Those who favor REC seek to explain skilled performance in terms of embodied activity that involves dynamic processes that span brain, body, and environment. Accordingly, cognitive processes are not, for example, conceived of as mechanisms that exist only inside individuals. Instead they are identified with nothing short of bouts of extensive, embodied activity that take the form of more or less successful organism-environment couplings. Likewise, embodied skills are acquired and emerge as a consequence of a history of interactions between learners and their embedding environments in ontogeny and phylogeny. Through sustained, context-sensitive, active engagements with worldly offerings, organisms are changed so as to be able, in Clark's (2015b, 5) apt formulation, to get "a grip on the patterns that matter for the interactions that matter."

In sum, through adjusting and attuning to the world over time, in complex and nested ways, organisms enactively evolve their most fundamental cognitive capacities. Their tendencies for interaction—their patterns of sensitivity and responsiveness—alter across many and varied spatial and temporal scales at once. Attention is a pivotal driver of these processes of adjusted sensitivity and refined responsiveness. The features of their environments that organisms attend to, and more fundamentally, how they attend in full-bodied active ways shift over time, leading to a variety of structural changes. How organisms respond to a range of worldly items thereby constrains their further interactions and creates adaptive changes—neural, bodily, and environmental—in turn altering organismic capacities and dispositions.

Through these processes of organism-environment adjustment the weights of neural connections change and are recalibrated. But relevant changes can also be located in the environment, such as when modification to the spatial arrangement of objects or created artifacts alters how organisms tend to interact with aspects of their worlds.

As just noted, REC acknowledges that cognitive capacities at least in part depend on structural changes inside organisms. But REC is cautious in the way it understands the character and basis of such changes—in particular, it steers clear of casting them in information processing and representational terms. Does REC's restraint on this front put it at odds with standard explanations of how neural changes in memory are wrought by the most basic forms of learning? The simple answer is: yes and no.

Continuity and Break

Can an adequate account be given in REC terms of the neural, biochemical, and genetic changes that contribute to the most elementary forms of learning and memory? At first glance, there seems no difficulty here if we attend only to the big lessons of the current state of the art. Consider Eric Kandel's Nobel Prize–winning research on learning and memory. In seeking to understand how learning produces changes in the human brain, Kandel focuses on simple animal models that he assumes obey the same basic principles of neuronal organization that apply to humans. This choice is justified to the extent that "elementary forms of learning are common to all animals with an evolved nervous system" (Kandel 2001, 1031; Kandel 2009).[1]

The full story of the relevant neural changes connected with learning, even in the simplest creatures, is enormously complicated. Telling it would involve a foray into a great many intricate details, not only in neuroscience but also in biochemistry and genetics. For our purposes it suffices to highlight two major lessons from Kandel's groundbreaking work. The first is that the special memory-enabling capacities of the hippocampus are not due to the intrinsic properties of its neurons. This is because all nerve cells have similar properties. What is important is "the pattern of functional interconnections of these cells, and how those interconnections are affected by learning" (Kandel 2001, 1030). The second lesson is that "learning results from changes in the strength of the synaptic connections between precisely interconnected cells … [such that] experience alters the strength and effectiveness of these preexisting chemical connections" (Kandel 2001, 1032). On the face of it, these lessons are entirely compatible with REC.

However, there's a catch. For in describing his work, Kandel (2001, 1030) tells us that "to tackle that problem we needed to know how sensory information about a learning task reaches the hippocampus and how information processed by the hippocampus influences behavioral output." This concern is perhaps to be expected, given that he set out with the express aim of contributing to a new synthesis that would combine "the mentalistic psychology of memory storage with the biology of neuronal signaling" (p. 1030). Such a goal is hardly surprising in today's research climate. Scientific research on memory is rife with casual use of the language of "memory traces," of "encoded and retrieved information," and of "the storage and retrieval of information and representations."

Despite the popularity of these familiar metaphors, close inspection of how they operate in science reveals them to have serious limitations—limitations that make them prime candidates for theoretical explication or elimination (see Roediger 1980). For example, as De Brigard (2014) observes,

> "Storing" is a rather misleading term. What seems to occur *when we encode information* is the strengthening of neural connections due to the co-activation of different regions of the brain, particularly in the sensory cortices, the medial temporal lobe, the superior parietal cortex, and the lateral prefrontal cortex. *During encoding*, each of these regions performs a different function depending on the moment in which *the information gets processed*. A memory trace is the dispositional property these regions have to re-activate, when triggered by the right cue, in roughly the same pattern of activation they underwent during *encoding*. (p. 169, emphasis added)[2]

REC assumes that De Brigard's analysis in the above quotation is mostly correct, *sans* its commitment to the highlighted claims that information is encoded and processed. REC seeks to explain basic forms of learning and memory entirely in terms of reenacted know-how.

Following Kandel's lead and focusing solely on the simpler kinds of procedural memory widespread in the animal kingdom remembering can be understood as the capacity to reenact embodied procedures, often prompted and supported by patterns of response that are triggered by external phenomena. Memory of this sort entails knowing what to do in familiar circumstances. It is surely not necessary to posit stored mental contents in order to explain the dispositional basis of such capacities (Ramsey 2007, chap. 5). The brain's underlying contribution to such capacities "turns out to be just a matter of either organizing extant synaptic circuits in new wiring patterns or switching on genes in neurons that produce new synapses. … The brain does everything without thinking *about* anything at all" (Rosenberg 2014b, 26–27, original emphasis).

Importantly, purely embodied know-how is not grounded in or mediated by any kind of knowledge; rather it can be understood as the overall responsiveness of a complex system laid down through habit and past experience (Barandiaran and Di Paolo 2014).

This REC account of basic memory can be provided, without gaps, so long as no appeal is made to the encoding and processing of information or representations. This crucial revision of the standard picture is pivotal for understanding what motivates the REC framework. According to the familiar cognitivist account, information is supposed to be picked up via the senses through multiple channels, encoded, and then further processed and integrated in various ways, allowing for its later retrieval.

The crucial question is: Does an information processing story add anything of explanatory value to the radically reductionist account of memory offered by Kandel, an account in terms of how experience modulates neural connections and weights? If not, then the information processing gloss is superfluous. But if the information processing account does add explanatory value, then exactly what additional contributions does it make, and precisely how do its explanations work?

There is a deep theoretical problem—which Hutto and Myin (2013) dub the Hard Problem of Content—that gives us reason to be skeptical of the standard information processing story. The problem arises from the fact that the notion of information that can be most easily called on to do serious explanatory work and provide the details of the information processing story is that of information-as-covariance. According to that notion, a state-of-affairs is said to carry information about another state-of-affairs if and only if it lawfully covaries with that other state-of-affairs, to some specified degree. The parade example is that of the age of a tree covarying with the number of its rings. Information in this sense is perfectly objective and utterly ubiquitous—it literally litters the streets. Moreover, this notion of information has impeccable naturalistic credentials: it is used in many sciences and thus can clearly serve the needs of a cognitive science with explanatorily naturalistic ambitions.

In trying to fill out the standard information processing story cognitivists face a dilemma. In dealing with the first horn of the dilemma, they can try to give a naturalistically respectable explanation of information encoding and processing by appeal to the notion of information-as-covariance. Yet if that is the only notion of information in play in cognitivist theorizing, then it is difficult to understand what it could possibly mean for information to be literally encoded. How can relations that hold between covarying states of affairs be literally "extracted" and "picked up" from the environment so as to be "encoded" within minds?

Perhaps it will be objected that what should be focused on here is not the medium but the message. Sometimes the information processing story is told in quasi-communicative terms of signaling and receiving messages. Yet how seriously should we take these analogies and the talk of encoding and decoding "messages"?

Again there are grounds for caution. Despite the widespread popularity of such talk, attempts to seriously explicate the nature of neural or mental "codes" and their alleged encoded content are few and far between. Goldman (2012, 73) gives a frank appraisal of the current situation: "There is no generally accepted treatment of what it is to be … a mental code, and little if anything has been written about the criteria of sameness or difference for such codes. Nonetheless, it's a very appealing idea, to which many cognitive scientists subscribe."

Moreover, how should we understand the nature and source of the putative contentful messages? The notion of information-as-covariance is surely not able to help us to understand how sense perception supplies the mind with messages—informational contents—that can be encoded and decoded, conveyed and communicated.[3] Again we lack any reputable scientific account of how to understand the idea that cognition is literally a matter of trafficking in such informational contents.

At this juncture it would be natural for cognitivists to try to instead deal with the second horn of the dilemma. They might try to call on some other naturalistically respectable notion of information that will enable them to tell their encoding and processing stories in full detail.[4] Telling a different tale requires identifying an alternative notion of information with sound naturalistic credentials that can do the additional explanatory work necessary to validate the standard information processing story. As things stand, it is at best unclear which of the competing notions of information has the right characteristics to play such a role.

In light of this analysis, it becomes clear that the "storage" metaphor is not the only, or even the most, problematic card in the cognitivist deck, *pace* De Brigard. The familiar cognitivist talk of information processing—certainly, to the extent that it takes seriously that information is some kind of commodity that carries abstract contentful messages—evokes equally serious scientific mysteries. Such mysteries need dispelling, one way or another—they want explaining or explaining away.

Less Can Be More

Famous cases in the history of science teach us that sometimes less is more. The need for and adequacy of explanantia are reassessed during conceptual revolutions in science as and when the nature of explananda is reconceived. The evolution of our thinking about what is required for explaining motion provides a shining example. Before a correct understanding of inertia was achieved in classical physics, bodies were thought to be in a natural state of rest. Because of this assumption the initiation and continued motion of objects were treated as primary explananda.

Explanations of the forces that moved objects from rest and kept them in motion were always assumed necessary, even in the basic cases. In Aristotelian and medieval frameworks, respectively, such explanations were given in terms of the medium or an impetus. Such proposals were plagued by internal theoretical problems. They also impeded progress by providing apparently compelling reasons to reject Copernican heliocentrism (Dijksterhuis 1961).

It was only with a major, hard-won shift in theoretical perspective, in which unaccelerated linear motion came to be understood as the natural state of objects, that it became clear that the problematic explanations previously sought were hollow and unnecessary. This is a clear instance in which, by reconceiving what needed to be explained, and removing the demand for distracting and misguided explanations, thinking was liberated and barriers to progress were removed.

Like those involved in the early debates about motion, cognitivists are prepared to overlook deep problems in their theorizing on the grounds that they believe the type of account they propose is the only kind that can meet the special explanatory needs of the case. Consequently, they hold that nonrepresentationalist accounts of intelligent behavior only appear credible if one systematically underestimates what the relevant explanations require.

Those persuaded by such reasoning hold that once the relevant explanatory needs are made transparent, it becomes evident that only posits of the sort cognitivists offer can properly account for the "stunning successes" of cognitive science (Shapiro 2014a, 214; see also Burge 2010). Prima facie, this assessment looks plausible enough given the current state of the art: "Examination of research on memory, attention, and problem solving … leaves little doubt that cognitive scientists are neck deep in representational commitments" (Shapiro 2014a, 218).

The master intuition behind these assessments of the special features of the explanandum is that cognizers must somehow manage to concretely represent abstract properties. Apparently, it is this explanatory need that rules out the possibility that intelligent behavior might be explained by contentless structural changes selected for through the history of an individual's interactions with worldly phenomena. Something more is needed, so the persistent intuition says, and that something must be a content derived from experiential encounters and carried by a discrete representational vehicle that modulates behavior.

The driving assumption is that abstract properties must be literally derived from a diverse array of perceptual inputs such that flexible responsiveness to such properties is controlled by discrete items that "stand in" for them. The key is that abstract properties need to be "re-coded into simple, useable objects" (Colombo 2014a, 16).

Clark and Toribio 1994 is the locus classicus for a contemporary version of the argument that the representation of abstract properties is needed to explain at least some cognitive tasks (see Degenaar and Myin 2014). Summing up the conclusion of that argument, positing representations is needed in order to understand how we manage to think about "states of affairs that are unified at some rather abstract level, but whose physical correlates have little in common" (Clark 1997, 167). To cognize in such ways requires "that all the various superficially different processes are first assimilated to a common inner state or process such that further processing can then be defined over the inner correlate: an inner item, pattern, or process whose content then corresponds to the abstract property" (Clark 1997, 167).[5]

Such distilled representational items are assumed to modulate a host of behavioral responses, as Matthen (2014) makes clear with his example of a dog that has learned to expect its food at 5 p.m. and thus has developed a single means to control a multiplicity of dispositions—sitting by its bowl, bothering its owner, whining, and so on—the manifestation of which depends on further contextual factors, such as whether its owner is present or absent. Matthen holds that "a single learned association controls all of these behaviors. This argues for discrete representation. The whole organism *cannot* just be rewired so that it incorporates all of these context-dependent behaviors" (121, emphasis added).

By this reasoning it can come to seem, by sheer intuitive insistence, that representations of some sort must feature in the best explanations of how cognizers "track abstract properties across situations" (Colombo 2014a, 16).[6]

The putative work that discrete, concrete representations mediating perception and action do is allegedly twofold: they are meant to distill what is encountered through learning and thereby they come to modulate a complex set of possible behaviors. Those who frame matters in this way will feel an irresistible need to posit such unifying, unitary representations. It can seem, given familiar lines of thought, that there must be concrete mediators of perception and behavior in the form of discrete representations that bear abstract contents.

The conclusion seems unavoidable. Crucially, however, this proposal raises precisely the concerns about the information processing story outlined above. The attempt to explain how information processing and contentful representations add explanatory value in such cases encounters the insuperable problems already mentioned. These problems are seemingly intractable when it comes to explaining in a naturalistically illuminating way how abstract information or content can be distilled from the world by means of environmental interactions and captured in concrete representational vehicles, without assuming some unexplained leap. Additionally—even if the above problem could be solved—we are owed an account of how the abstract content of a concrete vehicle could make any causal difference to cognition.

Is there another option? As Dreyfus (2014, 30) intimates, "In spite of the authority and influence of Plato and 2,000 years of philosophy ... one must be prepared to abandon the traditional view." In other words, we can escape the Hard Problem of Content by letting go of the idea that learning from specific cases requires interiorizing contents.

A Radical REConceiving

Conceptual revolutions are rare, to be sure. Even so, REC-style approaches—those that commit to the two central tenets discussed in chapter 1—have all the hallmarks of being bona fide revolutionary in that they press for a vision of cognition that conceives of it, at root, as a kind of embodied activity rather than continuing to think of cognition as essentially a matter of processing informational contents or the manipulation of contentful representations. The agenda for those in the REC family is not merely to tinker with and reform the notion of representation and thereby modestly adjust the vision of cognition associated with it. Instead RECers challenge the deep roots of the information processing paradigm that continues to dominate much of cognitive science. In this respect, they seek to replace that "whole system of concepts and rules by a new system" (Thagard 1992, 6). Thus, like other revolutions in thought, the REC movement does not merely require us to jump from one conceptual branch to another while staying within the same familiar tree; to adopt the REC framework requires switching to a new tree altogether.

In doing away with the idea that content is a defining feature of basic cognition, REC theories reject the most foundational notions of representational cognitive science. Going radical the REC way is to abandon the information processing and representationalist views of cognition in favor of a purely embodied know-how account. To embrace REC is to press for an extreme take on what Engel, Maye, Kurthen, and König (2013, 202) call the "pragmatic turn" in cognitive science—which is the intellectual movement "away from the traditional representation-centered framework towards a paradigm that focuses on understanding cognition as "enactive," as skillful activity that involves ongoing interaction with the external world." These authors suggest that the right move may be to replace the notion of representation rather than trying to reshape it. Of course, REC goes farther than just raising worries about the explanatory usefulness of invoking representations when it comes to understanding basic cognition; in arguing for the possibility that basic minds may be utterly contentless, it challenges not just representational theories of cognition but, more fundamentally, the information processing picture of cognition upon which classic cognitive science depends.

To get a sense of the magnitude of the shift in thinking that REC proposes, consider that to go radical in its way requires relinquishing the textbook content/vehicle distinction. Simply put, if basic minds lack content, then they lack vehicles that bear content. The link is clear, the notion of a vehicle—unlike that of, say, a neural structure—is logically dependent on the idea of contents that are carried or expressed by such vehicles. The content/vehicle distinction is a package deal that depends on accepting the very kinds of information processing, representational accounts of mind that REC challenges.[7] And when it comes to thinking about the architecture of cognition, abandoning the content/vehicle distinction is no small modification; it is not a minor tweak that leaves everything else intact.

In other writings, and most extensively in Hutto and Myin (2013), we have attempted to demonstrate that the version of the "pragmatic turn" in cognitive science advocated by REC is not only a live conceptual possibility but that there are powerful philosophical reasons for taking it seriously. As for its empirical credentials, as Engel, Maye, Kurthen, and König (2013, 203) highlight, a wealth of empirical findings that either demonstrates "the action-relatedness of sensory and cognitive processing or can be re-interpreted more parsimoniously in this new framework."

At this stage, making the pragmatic turn under the auspices of the REC movement is still to place one's bets on an emerging rather than established paradigm. Yet on the assumption that theory and practice are never truly isolated from one another, we can expect that making the theoretical head-shift that REC demands will yield significant changes to practice in cognitive science and other domains down the line.[8]

As a rule, rethinking a theoretical framework in fundamental respects yields changes to scientific and other practices—both in the way we conduct experiments and what we go on to do with findings. It would be jumping the gun to try to say in any interesting detail what the nature or extent of the changes would be if REC were more widely adopted: that is not clear at this stage of the game.

Even so, one immediate change—which we discuss further in the epilogue —is in sight already. Pretty clearly, if REC is adopted "our view of the brain and its function is likely to change profoundly. The conceptual premises of the pragmatic turn are likely to enforce a redefinition of basic neuroscientific explananda" (p. 207).

Revising our understanding of the brain's role in cognition and making other adjustments to theory and practice are to be expected because to embrace a truly revolutionary REC line is to join with those who "do not mean by 'cognition' what traditionalists have meant by 'cognition'" (Aizawa, 2014, 40). In assessing REC's revolutionary character, Aizawa observes that, unavoidably, "by adopting a new conception of cognition ... [revolutionaries] have detached themselves from the traditions of cognitive science" (p. 40).

Consequently, REC does "not so much solve traditional problems, as merely walks away from them" (Aizawa, 2014, 22). That is exactly right. Certain conceptual problems do not warrant straight solutions, they warrant dissolution by rethinking the underlying assumptions that bring them into being and make them seem, at once, intractable yet unavoidable.[9]

Walking away is precisely what REC recommends when it comes to dealing with impossible, framework-dependent questions such as how information can actually be acquired, stored, and processed in brains. Yet while this is so, Aizawa (2014, 21) overstates the case in claiming that this means RECers "use different tools to study different issues." It is clearly true that REC proposes a different framework for thinking about cognition and a different set of tools for studying cognitive phenomena than cognitivists employ. Nevertheless RECers target the same phenomena of interest to cognitivists—for example, perceiving, imagining, remembering—albeit by significantly REConceiving the nature of those phenomena.

In sum, REC avoids certain seemingly intractable theoretical difficulties by making a shift from conceiving the fundamental task of cognition as "accurately representing an environment to continuously engaging that environment with a body so as to stabilize appropriate coordinated patterns of behavior" (Beer 2000, 97). REC differs from its close conservative cousin, sensorimotor enactivism, in making a firm move away from all forms of representational thinking about basic cognition. RECers deny that all cognition, and in particular its root forms, involve information processing and contentful representations.

Importantly, REC's decision to go radical is motivated by reconsidering explanatory needs, especially in light of its diagnosis of the deep theoretical difficulties associated with the Hard Problem of Content. REC—at least as advanced in Hutto and

Myin 2013— thereby supplies something of special importance to those who want to break with cognitivism.

According to Aizawa (2014, 2015), most defenders of REC do not do enough to motivate its acceptance. He finds only a dearth of arguments. Aizawa (2015, 762) readily admits that "if one understands 'cognitive processes' as behavioral processes, then of course, 'cognitive processes' are typically realized in the brain, body, and world." But those who make this claim typically provide no argument in support of this identification.

Aizawa (2015) observes, for example, that Maturana offers only speculative pronouncements in "an oracular tone" (p. 760), and that Chemero too gives "no reason in support of his revisionary interpretation. ... There really is no argument there" (p. 764). Aizawa (2014) is certainly correct to say that "one *cannot argue* that cognition is embodied and extended, by observing that behavior is embodied or extended" (p. 40, emphasis added).

Of course the "no arguments for REC" claim is exaggerated. For example, one can find plenty of arguments over the years in the collected work of Dreyfus (2014), such as the infamous frame or infinite-regress arguments invoked there. Nor is it true that all defenders of REC have failed to note that "cognition has generally been proposed to be a cause of behavior ... [and failed to provide] reasons or evidence against it" (Aizawa 2014, 759). Concerns about the causal impotency of content are well known and have been used by RECers to motivate a shift in our thinking (see, e.g., Hutto 2013b).

Still, as arguments go, the Hard Problem of Content (see pp. 29–31) provides the basis for an especially tough one. If not by endorsing REC, how else might cognitivists handle it?

Handling the Hard Problem

Taking the radical REC line is motivated by a desire to provide a complete and gapless naturalistic account of cognition, right here, right now. REC predicts that, when combined with other E-resources, scientifically respectable contentless notions of information-as-covariance and the norms of biological functionality offer all that is needed for understanding basic minds.

All explanatory naturalists competing to understand basic cognition must ultimately face up to the Hard Problem of Content (HPC) one way or another. As noted, a straight solution to the HPC requires explaining how it is possible to get from informational foundations that are noncontentful to a theory of mental content using only the resources of a respectable explanatory naturalism. Adequate explanations of this kind have systematically eluded us.

Unlike phenomenal consciousness, it used to be said that mental content neither posed any "deep metaphysical enigmas" (Chalmers 1996, 24) nor "any deep philosophical difficulty" (Strawson 1994, 44). For the longest time such was the orthodox view of analytic philosophers of mind. Indeed, in the 1980s and 1990s not only did it seem possible that a workable naturalized theory of mental content might be in the cards, it looked to be just around the corner. In those days the race was on to be the first to cross the finish line. Hence, as Kriegel (2011, 3) observes, "A generation ago … finding a place for intentionality in the natural order—'naturalizing intentionality'—consumed more intellectual energy than virtually any other issue in philosophy." Nowadays, however, that research program has all but petered out.

Of course, a putting down of tools is what we should expect if someone had already crossed the finish line—namely, if a workable naturalized theory of content were currently in hand. Some believe this is just what happened. For example, Miłkowski (2015) tells us that the HPC "has already been solved" (p. 74); that it was "solved a couple of dozen years ago" (p. 74); that it was "already solved by Dretske (and Millikan, and Fodor, and Bickhard, by the way)" (p. 78); it was solved "at least in principle ... [by] ... various accounts ... [that] usually recruit a similar solution" (p. 83). Like the French soldiers who taunt King Arthur's knights in *Monty Python and the Holy Grail*, some defenders of Castle Cognitivism clearly assert there is no need to seek a solution to the HPC because "we've already got one."

What was the solution? Obviously, it required relying on something more than the notion of information-as-covariance. Miłkowski (2015) holds that other scientifically respectable notions of information were successfully called on for understanding cognition, namely information-as-control and information-as-structural-similarity (p. 76). Such proposals, or something close enough, are found in O'Brien and Opie's (2015) attempt to conceive of mental representation in terms of structural similarity. Those authors argue that the content of an analog representational vehicle is nothing but the structural resemblance holding between that vehicle and its object. Yet as Miłkowski (2015) acknowledges, rightly in our view, even if these richer notions of information and information processing do play a role in helping us to understand cognition, they do not constitute content; rather they are "necessary but not sufficient for representation" (p. 82). So, appeal to such notions alone cannot be the solution to the HPC.

The crowning move that allegedly solved the HPC was the invocation of the notion of teleological function by the likes of Dretske and Millikan (Miłkowski 2015, 83). Allegedly, that move is ultimately what accounts for the special kind of normativity needed to understand content, and it does so by calling on a biological notion of function. Or rather, as Miłkowski (2015, 84) claims, quite ambiguously, something like this move is what has already solved the HPC, "one way or another."

Despite Miłkowski's (2015) confidence, there are serious reasons to doubt that teleosemantics did the required work. There is widespread consensus among those with remarkably diverse philosophical predilections and agendas that teleosemantics fails to deliver an adequate theory of content. The big problem is that even if it is allowed that biological function entails some kind of normativity—for example, such that it implies that organismic responses can be misaligned with respect to certain features of the world that they target—the kind of normativity supplied falls a good distance short of what is required to explain how an organism comes to have mental contents with specified truth conditions.

This verdict on the shortcomings of teleosemantic accounts is repeatedly voiced. We are warned that "evolution won't give you more intentionality than you pack into it" (Putnam 1992, 33); that there is a crucial distinction between "functioning properly (under the proper conditions) as an information carrier and getting things right (objective correctness or truth)" (Haugeland 1998, 309); that "natural selection does not care about truth; it cares about reproductive success" (Stich 1990, 62).

In sum, the problem with teleosemantics, as discussed at length in chapter 4 of Hutto and Myin 2013, is that it fails to account for intensionality (with an s), which is needed to explain the semantic content of mental representations. If biosemantic theories are to deliver their promised truth conditional theory of content, they must spell out what specifies how a given mental representation takes or represents the world to be, and where this is thought to involve representing a targeted object or situation under a particular description or mode of presentation. Lacking such an explanation, biosemantic theories are in no position to account for the disquotational features of truth conditional semantic content.

Referring to the timeworn example of the fly-chasing frog, Fodor (1990, 73) warned us long ago that: "Darwin doesn't care how you describe the intentional object of frog snaps ... Darwin cares how many flies you eat, but not what description you eat them under" (see also Fodor 2008a, Rosenberg 2013).

Advancing a teleosemantic theory of content, Cao (2012) puts her finger on a crucial problem when she tells us "the target of a competent receiver's action is the same chunk of the world (slightly later in time) as whatever evoked the signal, that chunk is just what the signal carries semantic information about. ... In the case of the frog's visual motion detector, "fly" or "moving black dot" could both be simultaneously acceptable descriptions of the signal's semantic content" (p. 54).

The bottom line worry is, however, precisely that either description would be equally adequate for capturing the putative semantic content of the frog's mental representation, and we have no principled way of choosing between them. Indeed, a disjunction of many candidate descriptions would be perfectly adequate to capture the imagined content. It seems that "no amount of environmental appropriateness of a neural state or its effects is fine-grained enough to give unique propositional content to the neural state" (Rosenberg 2014b, 26).

Nor does this kind of worry evaporate if we swap a notion of mental content cast in terms of truth conditions for a weaker notion of content cast in terms of accuracy or veridicality conditions. Such adjustments risk being merely nominal. Certainly, without a convicting analysis of how introducing these weaker semantic notions into the mix makes a material difference to the basic problem we can expect it to recur. Thus, for example, as Burge (2010, 303) reminds us "Evolution does not care about veridicality. It does not select for veridicality per se." The same goes for accuracy. Fundamentally, as Burge diagnoses it, what's wrong with any attempted teleosemantic solution to the HPC is that there is "a root mismatch between representational error and failure of biological function" (p. 301).

Returning to Kriegel's 2011 assessment it would seem, in this light, that the correct explanation for the abandonment of the search for the required naturalistic theory of content is gloomier and far from triumphalist. In fact, "The naturalizing intentionality research program bears all the hallmarks of a degenerating research program. … [It] has run up against principled obstacles it seems unable to surmount. Far from being technical, the problems just mentioned are fatal" (Kriegel 2011, 3–4).

Another way to avoid the HPC is simply to adopt a pluralist and metaphysically noncommittal, antirealist take on the nature of scientific explanation. Those who advocate going the antirealist way hold that,

Truth or existence is not a necessary condition for theoretical posits like representations to be legitimate epistemological/methodological tools. … So, even if … cognitive systems are not representational systems at all … representationalism can still be successfully defended, and "representation-talk" can still be justifiably preserved. (Colombo 2014b, 271)

Accordingly, even if a naturalistic account of mental content is never forthcoming—even if the HPC is never solved—"little hangs on the matter" (Colombo 2014b, 271). Indeed, on this view, nothing does. This can be seen from the fact that the antirealist approach is so liberal that it allows that—despite their logical incompatibility—cognitivist and REC explanations can both valuably illuminate the very same phenomena, at the same time.

Sprevak's (2013) observations about fictionalism reveal why antirealist ways of avoiding the HPC are generally unattractive. Sprevak notes that going fictionalist is a way of avoiding having to make the unpalatable choice of either dealing with the HPC or going radical, à la REC, and thus revising the theory and practice of cognitive science. In this respect going fictionalist, prima facie, "offers a neat way out" (Sprevak 2013, 540).[10]

But there is a heavy price to pay. Fictionalists, like all antirealists, break the links between truth, existence, and explanation in ways that make it unclear just what kind of explanatory value is yielded by posting theoretical entities. Fictionalists, for example, must hold that a revealing explanation "can be provided by a fiction just as well as by truth" (Sprevak 2013, 556). Fair enough. Still, defenders of this family of views owe us an account of the precise nature and value of these explanatory offerings.

In a nutshell, the challenge for fictionalists and antirealists, more generally, is to say in exactly what way invoking fictions in such cases could possibly help. How can doing so provide any genuine illumination? Some possibilities can be ruled out in advance. In general, antirealist explanations cannot be causal in character. For example, if mental content does not really exist it cannot feature in causal explanations because only real entities can be causes. Although deciding on the correct analysis of causation and causal explanation remains a philosophically vexed matter, one thing everyone can agree on is that for something to be a cause it must exist. This presents a particular problem for fictionalist accounts of content because it seems that contents must exist and make a real difference if they are to ground relevant fictions about content. If fictionalists try to deny the grounding relation they are left with the problem of explaining how there can be fictions at all. This problem is quite serious on the assumption that fictions must be contentful (see Sprevak 2013, 553).[11]

The foregoing analysis reveals that antirealism will only provide a tenable way of dealing with the HPC if we are provided with an independently compelling account of how antirealist explanations explain.[12] Here we must be alive to the concern that "it is far from clear that such a have-your-cake-and-eat-it-too response is coherent" (Horwich 2012, 14).

In any case RECers agree with their realist-minded cognitivist opponents that questions of metaphysics matter in science. Thus, in rejecting an overly liberal pluralism, Fodor (2008b, 12) states: "There are lots of issues that a sufficiently shameless philosopher of mind can contrive to have both ways but not the issue between [REC] pragmatists and [cognitivist] Cartesians."

So how do realist defenders of cognitivism hope to handle the HPC? Either they admit that the problem is real and serious but adopt an optimistic attitude about the prospects for closing the explanatory gap in due course or they simply deny that closing that gap is necessary. Optimistic realists of the first sort assume the HPC will solve itself, that the metaphysics will come out in the wash. Optimistic realists of the second sort hold that the naturalness of content can be taken for granted even if no solution to the HPC is forthcoming. It is possible to discern these two versions of the optimistic realist stance—one more agnostic than the other—in Shapiro's (2014a) remark that:

perhaps Hutto and Myin are correct that *no* extant theory of content succeeds ...

[1] but this fact does not preclude future success. Why not take the challenges to naturalized theories of content that they pose to be a call for further work on naturalizing content, or further reflection on what it means for content to be natural? ...

[2] Perhaps one should look at struggles to naturalize content as misguided from the start, and see the tremendous gains of cognitive science as themselves sufficient to establish that content—whatever it is—is natural. (p. 218, original emphasis)

The first option requires that the HPC be solved in due course. The assumption is that mental representations exist and their contentful properties are what really and truly explain the successes of cognitive science, and that we will come to know more about how all this works someday. Assuming that content will one day be shown to be in good metaphysical order, optimists of this sort say, "Find the right architecture of mind, and the naturalizing strategy will follow. ... This is why ... arguments about naturalizing content are almost completely irrelevant at this stage of the inquiry" (Matthen 2014, 126).[13]

Things might turn out just as these optimists hope. If so, all will be well and good for representationalism. But until a solution to the HPC is firmly in hand, we can't know whether the metaphysics of content is in order. Consequently, we can't know now that contentful properties explain the successes of cognitive science. The history of science is littered with cases in which theoretical posits that were deemed to successfully explain some phenomenon were eventually eliminated from the science, and new explanations proffered. The simple fact is that despite some very confident statements to the contrary, no one is currently in the epistemically privileged position to know with any justifiable confidence that contentful representations will likely feature in ground floor explanations of the successes of cognitive science, let alone that they must do so.

The foregoing analysis reveals that, as things stand, cognitivist theorizing incurs heavy debts against the future. What we can know, right now, is that any current explanatory power cognitivist theories actually have is mortgaged against future theoretical assessments and developments that may not materialize. Optimistic cognitivist realists are placing a bet—a bet on which bookmakers would be forgiven for offering only long odds, given the long history of unsuccessful attempts to naturalize content and the growing skepticism about the explanatory credentials of content-involving mental representations in accounting for much intelligent activity.

Optimistic realists of the second sort do not require the HPC to be solved, ever—at least not by us. Instead they take it as an article of metaphysical faith that there is no fundamental tension in content being natural and our never managing to figure out how it could be so. Going this way would be to adopt a kind of mysterianism about content. Content mysterianism parallels mysterianism about phenomenal consciousness (see McGinn 1991). In general, mysterians allow that a given phenomenon can be natural even if we are cognitively closed to understanding how it could be so. Mysterians do not deny the existence of an explanation that fully accounts for how the phenomenon in question is wholly natural; they simply deny that we will ever have access to such an explanation. In such cases the truth is out there, but it will always elude us.

The trouble with the mysterian line is that it makes it difficult to rationally motivate realism about content. How can we be confident that mental content actually plays a part in the stunning successes of cognitive science if we are—*forever*—debarred from understanding how content could play a part in securing those successes?

Another option is out-and-out, utterly barren content eliminativism. Flatly denying the existence of all intentionality—both of contentful intensional (with-an-s) sort and any more basic variety—is, for true nihilists, the only rational conclusion to be drawn from "all the unsuccessful programs of research [that failed] … to provide a non-circular, let alone a naturalistic account of content" (Rosenberg 2015, 456).

The scorched-earth approach of such thoroughgoing eliminativism generates its own deep questions. Indeed, it leaves us trading one mystery for another. Anyone who claims that cognition is entirely a matter of contentless computations—for example anyone who allows that content falls out of the equation entirely, and offers no successor notion—will be unable to explain how organisms relate to and connect with targeted aspects of their worldly environments. Any theory of this extremely austere sort will be woefully ill-equipped to explain the array of findings that give us reason to think that cognitive activity is deeply influenced by E-factors.

This may explain why there have been so few advocates of content-free or computation-only accounts of cognition (Stich 1983; Piccinini 2008). Even computationalists who are skeptical about the existence of a deep metaphysical link between computation and content tend to advocate a more subtle position. They allow that even if computations are not essentially individuated by semantic properties—even if computations have a wholly nonsemantic and mechanistic nature—they can still be sensitive to semantic properties (Rescorla 2012a, 2014; Piccinini 2015).

Why so? The reason this is the preferred view is clear enough. Such theorists feel compelled to assume that "there are ... semantic properties that relate many computing systems to their environment—they relate internal representations to their referents—and are encoded in the vehicles in such a way that they make a difference to the system's computation, as when an embedded system computes over vehicles that represent its environment" (Piccinini 2015, 32). We have been at pains to show that paying for that assumption requires facing up to the HPC in one of the ways described above.

REC offers a different way out. It avoids the HPC by promoting a revolutionary shift in standard thinking about mind and cognition. REC's rejection of the I-conception of mind, but also the twin representational and computational pillars of cognitivism, is motivated by the avoidance of deep theoretical mysteries. But avoiding those mysteries requires more than just tinkering at the edges of cognitivist thinking about the basic character and architecture of mind or shifting our views on how cognitive processes function, interact, and unfold. REC asks us to REConceive and RECast our understanding of what cognition is, of how it works, and of what it does. It asks us to fundamentally adjust how we think about minds.

REC questions whether, on close inspection, there is a need to posit any kind of content at the basement level of cognition in order for the sciences of the mind to do their fundamental explanatory work. On the positive side, REC recommends getting by with something less—an alternative, contentless notion of intentionality (see Hutto 2008, chap. 3; Hutto 2011; Hutto and Myin 2013; and chapter 5 for more on this teleosemiotic replacement notion). In short REC avoids a host of intractable problems—most prominently the HPC and the problem of mental causation—by sticking with the idea that organisms target chunks of the world without assuming semantic contents make any causal or other explanatory contribution when it comes to saying how such targeting is possible.

Consequently, REC's strategy must not be confused with more extreme eliminativist ways of dealing with the HPC. REC does not propose simply biting the bullet and surrendering the idea that content is needed when it comes to understanding cognition—not even basic cognition—while offering nothing in its place.

Importantly, as with other major conceptual revolutions, should we succeed in radicalizing our conception of cognition and avoid what has been a theoretical hindrance in the old framework, we can still retain what is of value from that tradition by RECtifying it. RECtifying our thinking about the basic nature of cognition in this way will assist in unifying what is best in cognitivism and other E-approaches within a single framework. The hope is that pursuing such a positive program of work will yield many hard-to-predict philosophical, scientific and practical fruits down the line.

Some, like Shapiro (2014a), doubt that surrendering cognitivist thinking about basic minds is a tenable option for cognitive science. In particular, he advises against letting go of notions of information processing and representational content. For him, REC goes too far: he doubts that its thoroughgoing radicalism will take our thinking about cognition "to the next step" (p. 215). The jury is still out on that question. But given its revolutionary ambitions, Shapiro is certainly correct to say that REC is not simply "a more ferocious breed" of E-account, but "a different animal altogether" (p. 215).

Yet for all its ferocious radicality, it is important to recognize that REC is a package deal that seeks to accentuate the positive and eliminate the negative without messing with what's in between. In this respect REC's avoidance of the HPC is not as extreme as nihilistic eliminativist theories that try to handle it by giving up any idea of intentionality altogether.

3 From Revolution to Evolution

You say you got a real solution
Well, you know.
We'd all want to see the plan ...
—The Beatles, "Revolution"

REC's Positive Program

REC aims to show that basic cognition can be accounted for—in full and without explanatory residue—by understanding it in terms of thoroughly relational, interactive, dynamically engaged, world-relating activity—activity that does not involve any kind of information processing or the manipulation of representational contents.

REC's proposed adjustment to standard thinking about cognition strikes many as wholly negative. A persistent criticism of REC is that it only takes away and offers no positive alternative account in place of what it removes. As such, it is accused of doing "little to account for the stunning successes of cognitive science" (Shapiro 2014a, 214). This complaint is misguided on two fronts if it proves true that (1) the explanatory needs in question are only imagined and not real needs, and (2) the answers REC's rivals offer do not in fact adequately address those needs but only generate theoretical mysteries.

REC's positive contribution comes, not in the form of attempting to answer impossible problems, but by providing analyses and arguments designed to purify, strengthen, and unify existing representational and antirepresentational offerings. It aims to achieve this by RECtifying and radicalizing existing approaches to cognition—on both sides of the representationalist and nonrepresentationalist divide—through a process of philosophical clarification. Such work opens the door to pooling the most powerful resources and explanatory tools for understanding cognition from a wide range of sources under a REC banner.

Such work badly needs doing in the face of today's "rather bewildering variety of attempts to transform the principles of cognitive explanation" (Roy 2015, 92). As Roy notes, in the E-cognition family of views we are confronted by an embarrassment of diverse riches within which "one finds a minimal theoretical unity" (p. 92).

In seeking to address this lack of unity, the positive program REC offers is akin to what Clark (2016) hopes to provide with his Predictive Processing account of Cognition, hereafter PPC. What he offers is something special: not "yet another 'new science of the mind' but something potentially rather better" (p. 10). That "something better" is a way of understanding cognition that provides

a meeting point for the best of many previous approaches, combining elements from work in connectionism and artificial neural networks, contemporary cognitive and computational neuroscience, Bayesian approaches to dealing with evidence and uncertainty, robotics, self-organization, and the study of the embodied, environmentally situated mind. (p. 10)

In precisely the same spirit of unification, but with a different overall vision, REC seeks to unify what is best in today's sciences of the mind by clarifying the true character of the resources of existing theories of cognition. In keeping what is explanatorily best and leaving aside what is theoretically problematic, a process of RECtifying opens the door to forging productive alliances between approaches that would otherwise be in theoretical tension. The end result will be a stronger and more coherent understanding of cognition.

What does RECtifying look like in action? A clear example is the proposed adjustments REC would make to Clark's own attempt to unify theories of cognition through the lens of PPC.

A Certain Take on Predictive Processing

The Predictive Processing account of Cognition, PPC—at least in the general format in which Clark presents it—is clearly a worthy prize. It offers "an attractive 'cognitive package deal' in which perception, understanding, dreaming, memory and imagination may all emerge as variant expressions of the same underlying mechanistic ploy" (Clark 2016, 107).[1] Moreover, PPC is causing a real stir in philosophy, psychology, and neuroscience because of its apparently well-established scientific credentials through its links to Bayesian approaches to neuroscience (Clark 2013a, 2013b; Hohwy 2013, 2014; Friston 2010; Friston and Stephan 2007).[2]

There is much that REC and PPC agree about. That is hardly surprising since like REC, PPC—at its core—conceives of minds in essentially action-oriented terms. According to PPC, brains do not sit back and receive information from the world, form truth evaluable representations of it, and only then work out and implement action plans. Instead, tirelessly and proactively, our brains are forever trying to look ahead to ensure we have an adequate practical grip on the world in the here and now.

Focused primarily on action and intervention, the basic work of brains is to make the best possible predictions about what the world is throwing at us. Their job is to aid the organisms they inhabit, by being sensitive to the regularities of the situations those organisms inhabit. Brains achieve this by driving embodied activity that is dynamically and interactively bound up with the causal structure of the world on multiple spatial and temporal scales.

The exciting conceit of PPC is that it conceives of cognition in essentially anticipatory fashion, offering a dramatic reversal of the classical—sense-model-act—way of understanding minds. Crucially, for PPC, the brain is always poised and prepared to act, often readying multiple ways of doing so at once: "the human brain does not wait" (Clark 2016, p. 179, 145).

This goes directly against the idea that minds must first collect data from the world and only then build models and representations of it so as to act on the world. In a complete reversal of that familiar formula, the big idea of PPC, as illustrated by its rethinking of what it is to perceive, is that we are always trying to be cognitively ahead of the curve; indeed, that "we see the world by ... guessing the world" (Clark 2016, p. 5).

PPC sees the core business of cognition as that of making delicately balanced proactive, probabilistic, Bayesian predictions about what will be the most likely sensory perturbations. The ceaseless cascade of multilevel, multilayered cortical processing is all part of a singular brainy effort to predict sensory deliverances (Clark 2016, p. 146, 166, 167). Prediction error occurs when there is a mismatch between what brains predict and what is supplied to them by the senses. The brain's aim is to sensitively minimize the divergences between what it anticipates and actual sensory deliverances. This can be achieved either by making better guesses or acting in ways that will generate more fitting sensory deliverances. Either strategy can be used to reduce prediction error and uncertainty.

Nor is the brain a blunt instrument. In conducting its work, it toggles attention, in line with background knowledge about specific contexts of engagement, as a part of a meta-cognitive strategy for appropriately adjusting the weighting given to either sensory deliverances or top-down expectations, case by case (Clark 2016, 62, 64).

By PPC's lights, the brain's guesses are not a matter of constructing internal models of the world that are built upon passively received information furnished by our feature-detecting senses. Clark (2016) outlines a number of problems, limitations and questions that can be raised about the empirical adequacy of that rival, classical cognitivist view of cognition. As he sees it, the chief reason to steer clear of that traditional view is its failure to explain how we can "respond fluently to unfolding—and potentially rapidly changing—situations" (2016, 176).

In its bid to build a better mousetrap, PPC turns its back on the venerable cogntivist tradition and abandons the "last vestiges of the input-output model" (Clark 2016, 139). In this respect PPC makes another fundamental break with the classical cognitivist tradition. For under PPC's auspices there is no bright line between perception and the control of action (Clark 2016, 123). Consequently, many disparate elements of the old vision must be recast in the new PPC story that conceives of the brain as a singular predictive mechanism in which the main flow of information is top down. The forward flow of information in the classical cognitivist story is replaced by the forward flow of prediction error in PPC. Motor control is understood as top-down sensory prediction. Motor commands and efference copies are replaced by top-down predictions, and cost functions—also thought to be needed for motor control—are absorbed into predictions (Clark 2016, 117, 127, 131).

In turning the standard cognitivist vision of cognition on its head Clark (2015a) speaks of PPC's radical conceptual inversions. Indeed, when reflecting on PPC's central proposal he asks us to stand back and "savor the radicalism" (p. 4).[3] PPC heralds a sea change in our thinking about cognition. For example, Clark (2015a) claims, "Predictive processing plausibly represents *the last and most radical step* in [the] retreat from the passive input-dominated view of neural processing" (p. 2, emphasis added). Dramatically, he observes that if PPC is along the right lines then "just about every detail of the passive forward-flowing model [as promoted by classical cognitivism] is false" (p. 2).

If PPC is right, cognition has a fundamentally "restless, proactive, hyperactive and loopy character" (Clark 2015a, 1–2). In this and many other ways, PPC is compatible with—and ought to be amenable to—REC. Indeed Clark (2016, 1) even goes so far as to bill PPC as "the perfect neuro-computational partner for recent work on the embodied mind."[4]

These points of overlap between PPC and enactive, embodied theories of cognition are understandable since both approaches treat basic cognition as fundamentally active, world-involving, and self-organizing. Still, the potential for a happy union is dashed for RECers, at least, so long as PPC is viewed from Clark's CIC perspective—for that take on PPC does not relinquish the assumption that cognition is ultimately grounded in information processing and the manipulation of representational contents. On this pivotal issue REC is firmly at odds with cognitivist versions of PPC. Put otherwise, when donning a cognitivist guise PPC is fundamentally incompatible with REC. If PPC must don that guise then REC and PPC must part company.

Is cognitivism an adjustable feature of how we might understand PPC? Might PPC and REC be reconciled, after all? Clark does not think so: for despite recognizing the value of making strong connections with enactivist thinking, even at the very heart of the PPC story, he has difficulty seeing how the story of PPC can be told "in entirely non-representational terms" (Clark 2015b, 5).[5]

Certainly, on the face of it, Clark's formulation of PPC is replete with cognitivist terminology. He tells us, for example, that the brain's "predictions aim to construct the incoming sensory signal from the top down using *stored knowledge* about interacting distal causes" (Clark 2016, 6, emphasis added; see also p. 2); that brains are "knowledgeable consumers" (p. 6); and that as knowledgeable consumers, they are able to use generative models that seek "to capture the ways lower level visual patterns are generated by an *inferred* interacting web of distal causes" (p. 21, emphasis added). Putting all of this together, Clark says that in this process, the brain creates a "generative model that combines *top-level representations* ... with multiple intermediate layers" to self-generate data "using *stored knowledge*" in order to perceive a "meaningful structured scene" (p. 21, emphasis added). The end result is an intellectualized enactivism.

Looking from the top down, we can ask what—in Clark's formulation of PPC—the contents, not vehicles, are of the many states that he hypothesizes the brain is using in its probabilistic models. His answer is decisive enough: "It is ... precision-weight estimates ... *that drive action*. ... Such looping complexities ... make it even harder (perhaps impossible) adequately to capture the contents or the cognitive roles of many key inner states and processes using the vocabulary of ordinary daily speech" (Clark 2015b, 5, emphasis added).

Yet, despite telling us that it would be difficult to capture the contents used by the brain in ordinary language, Clark (2015a) offers an everyday analogy to provide an intuitive sense of the brain's situation when making active inferences. He asks us to imagine a game in which one participant attempts to describe what a second participant is seeing while the latter moves through a familiar environment—say, the living room of the first player's house. The catch is that the first player has no direct access to the visual scene and so can only make best guesses about it. The second player's role is to speak up and correct those guesses should they go awry, and to remain silent otherwise. Hence if the first player says, "There's a vase of yellow flowers on the table in front of you," the second player will either deny this claim or remain quiet.

Moving beyond analogy, how should we understand this at the level of theory according to cognitivist PPC? When it comes to understanding what is delivered by the senses (represented in the analogy by the second player's contribution), Clark (2015a) tells us that

in a very real sense, the prediction error signal is not a mere proxy for incoming sensory information—*it is sensory information*. ... Your "error signal" carried some quite specific information. ... *The content* might be glossed as "there is indeed a vase of flowers on the table in front of me but they are not yellow." This is *a pretty rich message*. Indeed, it does not (content-wise) seem different in kind to the downward-flowing predictions themselves. Prediction error signals are thus *richly informative*. (p. 5, emphasis added)

Sensory information in the form of newsworthy signals corrects our guesses, but since they only speak up when there is a mismatch there is a great saving on bandwidth economy and the possibility of cultivating a productive laziness (see Clark 2016, 26, 245).

At the very heart of PPC is the assumption that there can be "a mismatch between the sensory signals encountered and those predicted" (Clark 2016, 25). Things can also go well, as when the "downward predictions match the incoming sensory signal at many levels" (p. 27).[6]

The key point is that Clark tells the story of matches and mismatches in contentful terms. Thus he talks of responses that signal "error rather than well-guessed content" (Clark 2016, 43). His commitment to the idea that the brain's guesses are contentful is also clear when he says that "brains will not, of course, *get everything right*, all the time" (p. 8, emphasis added). Of course, the issue in the spotlight here—the one of interest to RECers—is not how frequently brains manage to get things right, but the very idea that they are able to get things right at all.

Despite pinning his cognitivist colors to the mast, Clark (2015b, 2016) still believes that PPC and enactivism can be brought together to mutually illuminate each other. Indeed, he thinks that a concessionary compromise on the part of cognitivism might lead to the end of the representation wars. He predicts peace in our time. Why so? Peace is allegedly in the cards because although PPC "openly trades in talk of inner models and representations, [it only] involves representations that are action-oriented through and through" (Clark 2015b, 4). Consequently, the representations of PPC "*aim to engage the world* rather than depict it in some action neutral fashion" (Clark 2015b, 4, emphasis added).

Hence, even though PPC is heavily committed to internal models, "instead of *simply describing* 'how the world is,' these models—even when considered at those 'higher,' more abstract levels—are geared to engaging those aspects of the world that matter to us" (Clark 2015b, 5, emphasis added). So, as Clark presents the situation, "What is on offer is thus just about maximally distant from a passive ('mirror of nature') story about the possible fit between model and world" (p. 4). This emphasis on the action-oriented character of representations deployed by the brain in Clark's formulation of PPC is his olive branch.

Clark's plan for peace clearly requires a bit of give-and-take from those in both the classical cognitivist and enactivist camps—resulting in an in-between, hybrid position. For him, such compromise is unavoidable: he can see no other way forward.

In contrast, RECers find Clark's proposed plan for collaboration unacceptable. This is because cognitivist PPC suffers from issues that can only be dealt with through RECtification. Such RECtification allows PPC to avoid intractable theoretical problems and results in a stronger, more positive, and potentially unifying view of basic cognition and the brain's role in enabling it.

Put otherwise, there is a fundamental problem with Clark's cognitivist formulation of PPC—one that ought to motivate a REC rendering of PPC. Clark (2016, 14) acknowledges the problem and construes it as follows: "How does the knowledge—the knowledge that powers the predictions that underlie perception and … action—arise in the first place?"

To fully understand the nature of the problem, it is important to recognize that cognitivist takes on PPC are strongly committed to the assumption that brains and scientists are engaged in essentially the same kind of intellectual work. This commitment is clear in Hohwy's (2014) account of how the brain deals with the problem of causal inference. He writes: "The problem of perception is the problem of using the effects—that is, the sensory data, that is all the brain has access to—to figure out the causes. It is then a problem of causal inference for the brain analogous in many respects to our everyday reasoning about cause and effect, and to the scientific methods of causal inference" (Hohwy 2013, 13).

Gerrans (2014) is even more forceful and direct in stating PPC's assumption: "a scientist explaining some discrepant evidence is *doing the same thing* as the oculomotor system controlling the trajectory of a limb" (pp. 46–47, emphasis added).

Recent research on brain reading, aka brain decoding, helps in getting a clear sense of how strongly cognitivists about PPC are committed to treating the work of brains and scientists as analogous. Using functional neuroimaging (fMRI), neuroscientists have tried to identify and decode the brain areas that show robust responses to visual orientation and motion (Kamitani and Tong 2005a, b; Norman, Polyn, Detre and Haxby, 2006).

The focus of these early studies was on the primary visual cortex, comprising V1 and extrastriate visual cortical areas, including V2, V3, V4, and V5/MT. Starting with the assumption that the job of these brain areas is to receive and transmit information, researchers set out to decode the information being processed by using fMRI activity patterns in the human visual cortex to sufficiently reliably predict what stimulus orientation a subject was viewing in individual trials. When extraneous factors were carefully controlled for, the findings were impressive.[7] The scientists were able to precisely decode which of a possible eight orientations a subject saw on individual stimulus trials by relying only on data about the fMRI activity in areas V1/V2.

Referring to later brain-decoding work by Reddy, Tsuchiya, and Serre (2010), Clark (2016, 95) tells us that "the experimenter is here in roughly the position of the biological brain itself. Her task—made possible by the powerful mathematical and statistical tools—is to take patterns of neural activation and, on that basis alone, infer properties of the stimulus."

It is important to be clear that fans of cognitivist PPC cannot and do not simply assume that the successes of brain decoding—predicting what subjects see on the basis of knowing what is going on in their brains—provide any insight into the information that the brain may itself be using in dealing with similar predictive tasks. Even if we allow, for the sake of argument, that brains are making informed inferences of a similar kind, "The ability to decode ... as an external observer does not mean that we have identified the code by which the brain represents ... information" (de-Wit et al. 2016). We must be on guard here precisely because "information in neuroscience is often measured with an implicit 'experimenter-as-receiver' assumption, rather than thinking in terms of 'cortex-as-receiver'" (p. 1).

Clark (2016) acknowledges this concern. He insists we must contrast "the perspective of an external observer of some system with that of the animal or system itself" (p. 15). The only proper way to do this would be to consider only the evidence available to the brain itself.

In trying to adopt the brain's perspective—and get a sense of its initial take on things—Clark (2016) says we must not assume that the brain knows anything about the possible worldly causes of sensory stimulation. Rather, "All that it 'knows' about in any direct sense, are the ways that its own states (e.g. spike trains) flow and alter" (p. 16).

Bootstrap Heaven or Hell?

The brain's task of guessing the world is, as Clark (2016, 16) readily admits, "a formidable task." It must try to optimize its inferences starting with only the extremely "slim pickings of the sensory evidence'" (p. 15). This is the task of generating optimal inferences about hidden causes starting with nothing more than the most minimal sensory data. Yet, according to cognitivist PPC, that "in essence, is the task of the biological brain" (p. 16).

To fully understand what is required for conducting the formidable task just described, it helps to consider the situation of another sort of scientist facing the same kind of inferential challenge. Martha Gibson (2004, 147) supplies us with a geologist's tale: she describes how a scientist might try to reconstruct a bit of geological history on the basis of knowing certain facts. In Gibson's example the geologist is concerned with a stratified rock formation that contains a layer of coal sitting above a layer of limestone. The layer of coal is treated as evidence of an ancient forest, the layer of limestone is treated as evidence of an ancient sea, and the fact that the coal stratum sits above that of the limestone is treated as evidence of their relative order of appearance—namely, as evidence that the sea was there before the forest. In making these successful inferences the geologist must have a grasp of the relevant facts and a means of representing them. Thus, she must have all the relevant concepts (e.g., to enable her to think about the coal stratum as an ancient forest, etc.) and background knowledge about the relevant relations (e.g., how stratified rock formations form over time) in order to make these judgments.

With this important comparison in place, we can ask afresh: Where does the brain get its conceptual resources and background knowledge so that it can represent information and make inferences?

Clark's (2016) answer is that it gets these resources by climbing—again and again—the stairway to bootstrap heaven.[8] It uses, so this story goes, the trick of making and correcting predictions to improve and refine what it thinks about hidden causes: it makes guesses that get corrected by the world. Such prediction-driven learning "exploits a rich, free and constantly available, bootstrap friendly teaching signal in the form of the ever changing sensory signal itself" (Clark 2016, 19). Thus, "the world can be relied upon to provide a training signal allowing us to compare current predictions with actual sensed patterns of energetic inputs. This allows well-understood learning algorithms to unearth rich information about the interacting external causes that are actually structuring the incoming signal" (p. 19).

Through repeating these steps the brain overcomes what "can seem an impossible task" (p. 20). It converts that seemingly impossible task into one that is merely formidable, and indeed into one that brains regularly solve in practice, simply by putting in the necessary time and effort.

In a nutshell, the above is Clark's story of bootstrap heaven. Yet there is every reason to think that the brain, as cognitivist PPC depicts its situation, in fact dwells in bootstrap hell.

Not enough consideration has been given to whether it is credible that information in the brain can be represented and used as any kind of evidence in the way adherents of cognitivist PPC propose in order to make their hypothesized visits to bootstrap heaven possible.

Consider Clark's (2016, 15) claim that "information here is used simply as description of energy transfer … [and that] talk of information … must ultimately be cashed out in terms of energies impinging on the sensory receptors." With this in mind, he tells us that "information talk, thus used, makes no assumptions concerning what the information is about" (p. 15). But that is already to assume too much. For if we take Clark's stated starting point seriously he is only entitled to the claim that "information talk, thus used, [can make] no assumptions [that] information is about [anything]" (Clark 2016, p. 15, with modifications).

This point matters when assessing Clark's claims about the brain's trick of distinguishing "what might be thought of as the mere transduction of energetic patterns via the senses from the kinds of rich, world-revealing perception that results … when and only when that transduction can be met with apt top down expectations" (p. 14). The problem is that, without solving the HPC or invoking magic, we have no idea how the cognitive fire gets lit in the first place.

It is one thing to start a large fire from a smaller one, even if under certain conditions doing so can be quite difficult. It is quite another thing to start a fire from scratch with only limited tools, of, say, flint and steel, especially when conditions are not favorable. But it really would be magic if one could start a fire with tools completely unsuited to the task, say a sponge and some wet cloth.

Fans of cognitivist PPC seem to be in the latter situation; their fundamental problem is that "networks of human brain cells are no more capable of [intrinsically] representing facts about the world … than are the neural ganglia of sea slugs!" (Rosenberg 2014b, 26).[9] At these crossroads, it seems there are three main positive ways forward:

1. Take the brain-scientist analogy seriously, give a straight solution to the HPC, and proceed to bootstrap heaven.

2. Don't take the brain-scientist analogy seriously, avoid the HPC, and go antirealist.

3. Go radical.

What exactly would it look like to go radical about PPC? How else, if not in terms of matching or mismatching contents, should we make sense of the way organisms reduce uncertainty—so-called error minimization—that lies at the heart of the PPC story? Reducing uncertainty and reducing free energy can be easily cashed out in REC terms. Recall that according to PPC, the brain is constantly seeking to minimize the degree of mismatch between internally generated sensory anticipations and what it senses in its encounters with the external environment. A REC take on PPC assumes that such embodied anticipations are grounded structurally and functionally in neural and other changes wrought through an organism's history of interactions (Byrge, Sporns, and Smith 2014; Bruineberg and Rietveld 2014). What we do and how we do it—what we experience—lead to changes, inter alia, in our neural setup, and hence in what we expect to experience.

Even the arch-cognitivist Hohwy (2014, 4), who understands PPC in Bayesian terms, admits that "priors are shaped through experience, development and evolution." Where Hohwy disagrees with REC is in his insistence that priors—the brain's initial expectation of possible ways the world may be—must be contentful attitudes instead of embodied anticipations.[10]

Having expectations about what we will experience sensorily need not be thought of as involving the making of any kind of contentful claim about the state of the world. Nor need we think of sensory perturbations that are involved in such matches and mismatches as supplying rich contentful messages that contradict the content of our expectations. Although the senses are sensitive to information in the environment, they can do their action-guiding work in a strong and silent manner (Travis 2004). There are ways of understanding the function of the senses in which representational contents play no part (Akins 1996). Yet even in such nonrepresentationalist construals, it is still possible to talk about what an organism expects to experience on some occasion as being in tension with—or failing to "match"—features of its current sensory experience.

This being so, our expectations can fail to match incoming sensory experience without this activity being construed as a content-based operation. This conclusion follows naturally if the senses do not have the job of telling us "how things stand objectively with the world" but rather of trying to ensure—within their suboptimal limits—that organismic activity satisfies specific, narcissistic organismic needs. Satisfying such needs surely involves being sensitive to and adjusting to the causal-probabilistic structure of the world, but such adjustments need not be representationally based and evidence driven. Consequently, in this analysis we can keep what is best in PPC while shelving the philosophically confounding talk of the brain making contentfully based "predictions," "inferences," and "hypotheses."

The best way to see how the REC proposal differs from Clark's (2016) formulation of PPC is to consider the crucial distinction he makes when he first introduces the latter—namely, the distinction he draws between two kinds of prediction, conscious and nonconscious:

> Prediction in its most familiar incarnation is something that a person engages in. ... Such predictions are informed, conscious guesses. ... But that kind of prediction, that kind of conscious guessing is not the kind that lies at the heart of the story I shall present. At the heart of that story is a different kind of prediction, a different kind of guessing. It is the kind of automatically deployed, deeply probabilistic, non-conscious guessing that occurs as part of the complex neural processing routines that underpin and unify perception and action. Prediction, in this latter sense, is something brains do. (p. 2)

REC proposes a quite different way of distinguishing these two kinds of cognition. It distinguishes not between conscious and nonconscious prediction, but between contentful predictions that can be made by persons with the right training and contentless embodied expectations of persons that brains play a major part in enabling.

Here is a piece of RECtified text, a heavily edited and modified version of Clark 2016, which captures the theoretical adjustments REC would make to Clark's PPC story about the crucial work done by brains in cognition:

> Nothing here requires [the brain] to engage in processes of ... [contentful] prediction or expectation. All that matters is that ... [its] systems be able to [anticipate and be adjusted by sensory perturbations] in ways that make the most of whatever regularities ... [to which it is attuned, because such attunement has] ... proven useful ... [in response to such regularities in the past]. (p. 27)

Although we will return again to REC's take on PPC in chapter 7, this puts us in a position to give at least a preliminary reply to Hohwy's (2014) "basically friendly challenge"—that of showing how enactivism "can *avoid an epistemic, inferential reading* in terms of the self-evidencing that entails an evidentiary boundary and thus decoupling" (p. 19).

Hohwy claims that the brain must confront a special epistemic problem because, unlike Clark, he holds that PPC paints "a picture of the brain as a secluded inference-machine" (p. 19). According to such a picture the mind-brain is forever secluded and cut off from knowledge of the world. This rendering of PPC assumes a "stark mind-world schism," one that entails global skepticism (p. 19).

Yet why assume that brains are in the epistemic predicament of being cut off from the world? The reasoning is both simple and familiar: it would be an "ideal but impossible design" for the brain to make any direct comparison between its internal estimates and "true states of affairs in the world" (Hohwy 2014, 4).

If the brain really were a representational device—one that only had access to its own contents—then it would be in no position to compare in any direct way what it represents as being the case with what really is the case. If we take this traditional problem of access seriously, the best such a representational brain can do is make inferences to the best explanations about how things might stand with the world. This is why, according to Hohwy (2014), even in optimal cases, the best the brain can do is hit on a hypothesis that best explains or explains away the occurrence of some evidence.

Of course, REC rejects this epistemic take on the brain's situation since it denies that brains are in the business of trading in contentful representations. Rejecting the epistemic reading allows REC to avoid having to deal with this old philosophical chestnut: If minds are in principle forever secluded from the world, how do they come by contents that refer to or are about the inaccessible hidden causes and external topics they putatively represent in the first place?

Clark claims that PPC is best understood in terms of brains using models that do more than just neutrally describe the world. But the above worries—about the origins of the putative contents that the brain allegedly treats as evidence—give us reason to doubt that the brain literally uses any contentful models at all. Of course, there is every reason to suppose that reliable correspondences hold between neural activity and aspects of the world. That is what the brain-decoding experiments reveal. Hence the brain might be used as a model by scientists in the sense that they could use brain activity to make reliable predictions and claims about how things stand with the world on the basis of their background knowledge. But if REC is right, that is not what the brain itself does in supporting basic cognition.

In sum, the REC take on PPC is that although we have ample reason to think that brains play a central role in enabling embodied expectations, we have no grounds to suppose that the brain does its important work by modeling or describing anything at all.

Showing that he is aware of this possibility Clark (2015b, 4) asks: "Why not simply ditch the talk of inner models and internal representations and stay on the true path of enactivist virtue?" Why not, indeed! Finding our way to that straight path is not a matter of brokering peace through theoretical compromise. Only genuine clarification can achieve true philosophical peace. But how will we know that we have clarified matters properly? Wittgenstein (1953, section 133) supplies the answer: "The real discovery ... is the one that gives philosophy peace, so that it is no longer tormented by questions which bring itself into question."

4   RECtifying and REConnecting

You tell me it's the institution

Well, you know
You better free your mind instead
—The Beatles, "Revolution"

RECtifying

There are many approaches to cognition other than PPC with which REC can naturally ally.[1] Two prominent approaches—Autopoietic-Adaptive Enactivism and Ecological Psychology—are considered in this chapter. The aim is to show that these approaches are better and stronger when partnered with REC, rather than some variant of cognitivism when it comes to thinking about the nature of basic minds.

Making Sense of Sense Making

Autopoietic-Adaptive Enactivism, or AAE, promotes a thoroughly biological vision of cognition grounded in the life-mind continuity thesis, which holds that "Mind ... reaches down to the simplest organisms. Life is always 'minded;' ... and the richer a living form, the richer its mind" (Colombetti 2014, p. xv). AAE conceives of mind and cognition as emerging from the self-organizing, self-creating, and self-preserving activities of a subset of living organisms—those that exhibit autonomy and agency, understood in a particular way.

AAE insists that organisms are not slavishly—or mindlessly—responsive to external factors; they do not respond to worldly phenomena in a fixed or blindly habitual way. Agents are shaped by habit but are autonomous in the sense that they always remain responsive to the particularities of their current contexts. They always exhibit some degree of flexible and spontaneous responsiveness such that their actions are not dictated in fully predetermined ways by the world.

Fans of AAE make much of the fact that organism-environment couplings can be more or less effective. They see this as implying the existence of a kind of biological normativity that goes beyond any norms that can be associated with mere autopoiesis. Arguably, the self-organizing, self-sustaining activities of all living beings, as exemplified by the metabolic self-production of single-cell organisms, entail the existence of very basic goals tied to maintaining a continuing identity.

It has long been recognized that any biological norms connected with autopoiesis are too open-ended to account for the way agents target and respond to specific features of their worlds: "self-constitution of an identity can ... provide us only with the most basic kind of norm, namely that all events are good for that identity as long as they do not destroy it" (Froese and Di Paolo 2011, 8).

Consequently, pure autopoietic versions of enactivism need augmenting in order to account for the sort of biological normativity that is a hallmark of even the most basic kind of cognition (Di Paolo 2005). Something stronger than autopoiesis is required in order to understand the normativity of cognition.

Enactivists of the AAE stripe think adaptivity can do that work. Adaptivity is the process through which certain sensorimotor loops of agent-environment couplings come to be favored as "more useful for the agent than others, such that those become more salient or meaningful" (Heras-Escribano, Noble, and de Pinedo 2015, 3). Proponents of AAE hold that "adaptive regulation is an achievement of the autonomous system's internally generated activity rather than merely something that is simply *undergone* by it" (Froese and Di Paolo 2011, 9, emphasis added). In sum, although autopoiesis provides a necessary foundation for agency, it is the capacity of agents to adapt selectively to specific features of their environment that gives real cognitive bite to autopoiesis.

AAE's great virtue is that it advocates thinking about the basic goal-directed cognition of agents in terms of biological norms without invoking any of the standard equipment that cognitivists insist is required for making sense of such phenomena—namely, mental contents, prior intentions, directions of fit, and the like.

In spelling out these ideas, proponents of AAE go further and characterize the sort of adaptive responding required for basic cognition as a kind of "sense making" (Weber and Varela 2002, but see also, e.g., Di Paolo 2009; Di Paolo, Rohde, and De Jaegher 2010; Colombetti 2014). Sense making is said to occur whenever an agent treats the perturbations it "encounters during its ongoing activity from *a perspective of significance* which is not intrinsic to the perturbations themselves" (Froese and Di Paolo 2011, 9, emphasis added). Sense making is characterized as a "*process of meaning generation* in relation to the *concerned perspective* of the autonomous system" (Froese and Di Paolo 2011, 7, emphasis added). Sense making is "the enaction of a *meaningful* world by an autonomous system" (Froese and Di Paolo 2011, 7, emphasis added, see also Colombetti 2014, 15).

Thompson and Stapleton (2009, 26) tell us that "even the simplest organisms regulate their interactions with the world in such a way that they transform the world into a place of salience, meaning, and value—into an environment (Umwelt) in the proper biological sense of the term. This transformation of the world into an environment happens through the organism's sense-making activity."

The so-called meaning generated through sense making is thus neither a feature of the external environment nor something internal to the agent. This is famously illustrated by the example of a bacterium engaging in sense making and thus enacting its world by responding in different orientations to a sugar gradient. Having the status of an affordance, the "sugar's edibility is not an intrinsic property: it is only valuable in relation to the agent that takes advantage of it" (Heras-Escribano, Noble, and de Pinedo 2015, 4; Thompson 2007, 125). Advocates of AAE go further, arguing that "if autopoiesis (or autonomy) suffices for generating a 'natural purpose' …, adaptivity reflects the organism's capability—necessary for sense making—of *evaluating the needs and expanding the means towards* that purpose" (Froese and Di Paolo 2011, 9, emphasis added).

There is no reason to deny cognitive status to noncontentful world-directed activities of living creatures, including ourselves, that are capable of detecting, flexibly tracking, and interacting with salient features of an environment in context-sensitive ways. REC and AAE agree that that sort of world involving activity does not suffice for the kind of intellectual sense making or meaningful evaluation that involves contentfully representing the world to be one way rather than another.

Put otherwise, REC and AAE agree that having a mind that can make sense of its world by referring to aspects of it and thinking about it in ways that can be correct or incorrect is not a basic biological endowment. Meaningfully evaluating situations in those sorts of contentful ways requires special forms of enculturated cognition: forms of cognition that only come into play for creatures that have mastered special normative practices, practices well beyond the reach of simple organisms.

The kinds of misalignments and failures of worldly engagement that can occur at the level of basic cognition do not involve making errors that are anything like errors of contentful judgment. Failures to engage with the world effectively are not, and are not explained by, failures to describe, depict, or say how things stand with the world.

Yet AAE goes beyond REC in holding that basic minds are capable of "sense making," "meaning generation," and "evaluating needs" in nonrepresentationalist ways—in ways that are unlike the ways implied by the usual connotations of these terms. This has raised questions about how exactly the central constructs of AAE—sense making and meaning generation—ought to be understood.

On this score, AAE has been accused of "containing a number of open problems and several insufficiently clear or underdeveloped ideas and concepts" (De Jesus 2016, 131). In a bid to place AAE on a sounder theoretical footing, De Jesus (2016, 131) advocates "non-anthropocentric reconceptualisation of the notion of cognition as the active and creative process of bio-semiosis by bio-semiotic systems." He rightly recognizes that in making sense of sense making it is crucial to avoid the mistake of "simply projecting the analogues of human experiences down the phylogenetic scale" (p. 134).

His proposed solution is to make sense of the notion of sense making through the lens of biosemiotics. It is claimed that this maneuver adds theoretical sophistication to AAE's foundational notion while crucially avoiding anthropocentrism and anthropomorphism. Yet, on De Jesus's rendering, taking this proposal seriously requires accepting that organisms are responsive to signs that stand for something else, even in the most basic acts of cognition.

On the face of it, De Jesus's (2016) proposal might look as if it could help to put AAE on a more secure theoretical footing. Crucially, he claims that a biosemiotic treatment of natural signs can avoid the twin problems he mentions and that it should not be confused with attempts to understand basic cognition in terms of mental representations. This is because a biosemiotic account of natural signs understands the way living systems respond to them to involve irreducible triadic processes in which A interprets B as "standing for" C (p. 137). Yet this attempt to provide a nonrepresentational account of sense making—by invoking the notion of natural signs that "stand for" worldly items—is entirely compatible with, and indeed, invites a Millikan-style analysis that is as full bodied as any representational theory can be. To take De Jesus's proposal seriously, at least in the unadulterated form just presented, would be self-defeating for AAE, since rather than providing a new and secure way of making sense of sense making, it entails that even the simplest acts of cognition involve sense making of a kind that advocates of AAE hold is only available to certain enculturated beings.

Like De Jesus, Clark too makes an attempt to provide new tools for understanding enactivist sense making. He proposes to do so in intellectualist terms as part of his attempt to unify various approaches to cognition under a cognitivist PPC banner. He claims that one of the attractive features of his PPC plan for peace—one that should be welcomed and embraced by enactivists—is that it bears the special gift of supplying the explanatory resources to "cash ... enactivist cheques" (2015b, 3; see also Clark 2016, 290).

Which "cheques" exactly? In particular he thinks that his PPC account can help to elucidate "the very real sense in which human agents help construct the very worlds they model and inhabit" (Clark 2016, 209). For him, PPC has the potential to demystify the otherwise "mysterious-sounding notion of 'enacting a world'" (p. 209). By implication he sees PPC as promising a way of making sense of sense making. In his view, the brain's hypothesized predictive modeling "apparatus delivers a firm and intuitive grip on the nature and possibility of meaning itself. For to be able to predict the play of sensory data at multiple spatial and temporal scales just is ... to encounter the world as a locus of meaning" (p. 3). Clark aims to make sense of enactivist talk of our constantly coconstructing our worlds in terms of representing the world by means of predictive modeling (p. 4). Yet as we have seen in the preceding chapter, the HPC is the Achilles heel of his PPC proposal, revealing it to be explanatorily hollow.

These attempts by De Jesus and Clark seek to intellectualize AAE in a bid to secure its theoretical foundations. In contrast, REC provides the right set of tools for making sense of sense making at the level of basic cognition without perverting its main messages. REC advocates holding on to the idea that the relevant notions of sense making and meaning are best understood in terms of "pure know-how learned through unprincipled interaction with the world. [In this sense, basic] cognition is primarily a relational form of meaningful engagement" (Cappuccio and Froese 2014, 4).

Keeping Affordances Affordable

Clark's (2016, 246) attempt at unification of cognitive science under the auspices of cognitivist PPC seeks to build bridges with and benefit from approaches in robotics that have rediscovered "many ideas explicit in the continuing tradition of J. J. Gibson and ecological psychology."

He lays great emphasis on the lessons to be learned from the classic example of how baseball outfielders position themselves to catch fly balls. Ecological theorists hold that rather than making inferentially based predictions involving complex calculations in the head, it appears that outfielders get where the ball is by running so as to keep its trajectory looking reliably straight, and "making moment-by-moment self-corrections that crucially involve the agent's own movements" (Clark 2016, 247). Outfielders apparently solve the problem of getting to the right place at the right time by being sensitive to environmental information in ways that enable them to adjust their embodied activity without having to form rich inner contentful models of the world.

We agree with Clark (2016) that the idea of a shared cognitive load is the main take-home lesson of developments in robotics—namely, that the coevolution of morphology and control has provided a "golden opportunity to spread the problem-solving load between brain, body and world" (p. 246).

The crucial point is that ecological psychology's way of characterizing acts of perception of this kind understands organism-environment coupling in quite a different way than that promoted by traditional cognitivist accounts. In Clark's (2016) CECish way of construing things, sensing is not all about "getting enough information inside ... so as to allow the reasoning system to 'throw away the world' [and] to solve the problem *wholly* internally" (p. 247, emphasis added). This is the main, but limited, lesson that he takes away from ecological psychology.

For Clark, the main strategy employed in situated acts of cognition—as in the outfielder case—is to use various sensory channels as "an open conduit allowing environmental magnitudes to exert a constant influence on behaviour" (p. 248). Seen from Clark's CEC standpoint, such coordinated sensorimotor engagement with the environment—the self-generated motor activity—acts as a "*complement* to neural information processing" (Lungarella and Sporns 2005, 25; Clark 2016, 248).

Crucially, the idea that *any* information is actually collected from the world—that any information somehow "gets inside" the perceiver's head and is neurally processed—is an in-between position that is very un-Gibsonian. Ecological psychology was inspired by the very attempt to provide explanatory tools for a thoroughly nonrepresentationalist cognitive science. It makes central use of the notion of direct perception of environmental affordances in ways that do not require the positing of inferences involving mental representations. Thus Clark's CECish attempt to incorporate Gibson's groundbreaking work in ecological psychology within a cognitivist PPC framework goes against the natural grain of the Gibsonian framework.

Classical ecological psychology regards perceiving as an active, relational phenomenon, where organisms are taken to be defined by sets of abilities and it is assumed that they occupy niches understood "as the set of situations in which one or more of [an organism's] abilities can be exercised" (Chemero 2009, 147–148).

Clark (2016) is not the first and certainly not the only philosopher to propose a cognitivist recasting of Gibsonian ideas (see also Siegel 2014; Hufendiek 2016). Millikan (2005, 174) makes an earlier attempt at such appropriation, telling us that "James J. Gibson did not advocate speculating about inner representations. Yet his notion that in perception we perceive certain affordances (opportunities for action) suggests that perceptual representations are [pushmi-pullyu] representations."

Pushmi-pullyu representations, according to Millikan (2005), are the most primitive representations. Still she understands the content of such representations to be their specified satisfaction conditions (see p. 171). In fact, Millikan holds that they have truth conditional content. So, despite being the most primitive kind of representations, pushmi-pullyu representations are also in this respect as representational as representations get.[2] This is why Millikan (1993, 12) holds that pushmi-pullyus are sufficient to ground "flatfooted correspondence views of representation and truth."

Yet she also notes that "the ability to store away information for which one has no immediate use ... is surely more advanced than the ability to use simple kinds of Pushmi-Pullyu Representations" (Millikan 2005, 175). Still, unlike REC, she takes it to be unproblematic that much cognition is about "collecting," "picking up," "applying," and "transmitting" natural information. In this vein, she claims:

We have thoughts of substances in order to be able to collect information about substances, which information we pick up on some occasions and apply on others. To pick up information about a substance you must be in a position to interact with the substance, other things that are influenced by the substance or that influence it. Natural information is transmitted in the causal order, and you have to be in the causal order, with whatever the information is information about to receive it. (Millikan 2005, 115)

Although Clark, Millikan and others set out explicitly to appropriate Gibsonian ideas within a cognitivist framework, there are others who invoke Gibson so as to actively resist the cognitivist framework—nonetheless they sometimes unwittingly make use of cognitivist ideas. Chemero's work is especially interesting in this regard. The primary aim of Chemero 2009 is to update ecological psychology, making it "dynamical root and branch" (p. 150). His ambition is to refine and augment ecological psychology's theoretical resources so that they can best serve experimental and explanatory needs.

As part of this effort, Chemero raises an important rallying cry. He recognizes the potential theoretical advantages of forging "the natural, but largely unmade connections between ecological psychology and ... the burgeoning enactivist movement in the cognitive sciences" (p. 152). Back in the day, he gave a sober assessment of the state of play, recognizing that "much more work is required to genuinely integrate ecological and enactive approaches" (p. 154).

With respect to at least some aspects of his own project of ecological dynamics, Chemero's (2009) assessment still holds true. Like REC, Chemero gives pride of place to organismic interactions and how they develop cognitive tendencies over time. However, a major obstacle to a thoroughgoing integration of Chemero's version of an ecological dynamical approach with

REC is that the former—despite itself—remains committed to the language, if not the framework, of information processing. Some of Chemero's ways of talking – when he speaks of the "provision," "use," "gathering," and "pickup" of information "about" affordances— are anathema to a nonrepresentational rendering of Gibson. Such talk (see Chemero 2009, 154–161) suggests an underlying commitment to an information processing story that is certainly inconsistent with REC.

RECtification of Chemero's (2009) account is therefore required if it is to be amalgamated into a larger REC framework that incorporates the best ideas from other, more thoroughly nonrepresentationalist, ecologically inspired accounts of cognition (Rietveld 2008; Rietveld and Kiverstein 2014; Kiverstein and Rietveld 2015).

As van Dijk, Withagen, and Bongers (2015) acknowledge, Chemero is not unique in generating a tension by explicitly endorsing nonrepresentationalism and conceiving of the way information is used in cognition in cognitivist terms. The same problem occurs in the writings of other prominent theorists within the ecological tradition. Indeed, van Dijk, Withagen, and Bongers (2015, 212) note that it in today's climate is "hard to get a contentless reading of even the most progressive ecological theories."

Van Dijk and colleagues (2015) propose a way out—arguing that a careful analysis of Gibson's corpus reveals that though "the notion of 'information pick-up'… takes on a content-carrying connotation in Gibson's early work, [it can be] understood in a content-less sense. Having the sensitivity, or the openness, to 'resonate' to the ambient patterns available, the animal picks up [on] those patterns as information for perceiving and acting. Such a reading, we feel, would give a fruitful and more charitable account of ecological theories" (p. 213).

Crucially, in true RECish spirit, they note "there need not be any content involved at all, as information for affordances cannot be evaluated as being more or less true or accurately corresponding to an affordance—there are no conditions to satisfy it being about the affordances. ... Information can be more or less useful for adapting to the environment, that is all" (p. 213).

These authors don't see providing a revised, properly nonrepresentational account as a problem. They propose that the idea of organisms responding to "information about" affordances and of "informational pickup" need not feature in the theoretical commitments of mature ecological accounts. Information processing and manipulation need not be working parts of the explanatory apparatus of such theories. Instead they propose, despite the fact that information processing talk is "sticky" in the sense of being hard to avoid, even for those who know better, it is best seen as nothing more than a kind of intellectual hangover. The unreflective use of information processing language is brought on by overexposure to a resilient intellectualist early modern picture of what minds are and how they become furnished through the senses. Yet, both the talk and the underlying theory are entirely revisable. Should this prove to be the case, then ecological accounts can be rendered wholly REC-friendly by refining their understanding of information and the role it plays in basic cognition, potentially yielding much theoretical clarity and explanatory gain.

Following REC's lead, van Dijk, Withagen, and Bongers (2015, 212) suggest that ecological psychologists need to embrace the notion of "information for" in the place of the notion of "information about." In line with that change, these authors propose that "information needs to be understood 'teleosemiotically.' … The high level of array-environment correspondence makes the patterning in the array very useful to the person. But only as these patterns are used, need they be considered information for perceiving or acting on affordances" (p. 213).

In sum, these high-profile examples of AAE and Ecological Dynamics both illustrate how, through RECish radicalizing, it is possible to ensure that powerful and promising explanatory resources can remain on a fully stable theoretical footing while warding off the threat of their incorporation into an unwanted and obfuscating cognitivism. This illustrates how REC's program can lead to a true "crystallization of enactivism" and, eventually, to the full and complete development of "a positive alternative to representationalism" (Heras-Escribano, Noble, and de Pinedo 2015, 1). The foregoing analysis makes clear how RECtification can elucidate and unite, after adjustment, the best of various cognitivist and nonrepresentational approaches to basic cognition within a single, secure and explanatorily powerful conceptual framework.

REConnecting

REC's story does not end here. All that has been said so far applies only to basic cognition, and—if REC is right—giving a full account of cognition requires giving an account of content-involving cognition as well. In challenging unrestricted CIC, RECers have focused on explicating the nature of basic, contentless minds, but telling the full tale of cognition entirely in such terms has never been REC's ambition. REC does not hold that cognition is always devoid of content. In failing to acknowledge this nuance, unrestricted-CIC critics systematically underestimate REC's ambitions and resources.

Aren't RECers out-and-out content deniers? Shapiro (2014a, 214) asserts so: "Hutto and Myin ... wish to deny that minds have any content. They deny what seems undeniable—that mental states, like my thought that Madison sits on an isthmus, and even perceptual states, such as my visual experience of sailboats on Lake Mendota, represent anything. But how can a thought that p not be about p? How can my perception of o not have the content o?"

To take Shapiro's claim seriously ought to induce wonder at the subtitle of this book: "Basic Minds Meet *Content*?" How can committed RECers be concerned with such a topic? Apparently our arguments for thinking that basic minds may be contentless have made it appear, to some, that RECers are extremist zealots, naysaying revolutionaries who's only slogan is "Just say no to content!"

This is a gross oversimplification and mischaracterization of the REC outlook. To say that basic minds lack content is not to say that all minds lack content. REC has always acknowledged the existence and importance of content.[3] What REC denies is unrestricted CIC: the thesis that there is content whenever and wherever there are basic minds, or putting it differently, that lack of content entails lack of mind.

To prevent misunderstanding, it is important to clarify what basic cognition means in a REC account. It is common in the scientific literature to treat "basic" as a designation that only picks out very low-grade forms of cognition. This is not the REC view. REC assumes that basic minds are contentless, and that they are the most fundamental kinds of minds—namely, that they are phylogenetically and ontogenetically basic. Even so, basic cognitive activity can be extremely flexible, open-ended, and context-sensitive.

What then is REC's take on content? Contents exist, though they aren't things. And basic minds can come into commerce with contents. This can happen in various ways. For example, basic minds can come into communion with and under the influence of the contentful minds of others. Think of how, say, dogs cooperate and think with us—see, for example, Merritt 2015. Again, let there be no mistake: REC's account of basic cognition not only concerns rudimentary forms of animal cognition, it also applies to central forms of human cognition, such as perceiving, imagining, and remembering in both children and adults.

What is contingently true of humans, though, is that through mastery of special, content-involving practices our basic minds can enter into more even-handed relations with the contentful minds of others. If such practices can be mastered, a basic mind will gain new, content-involving repertoires of its own.

From the REC perspective content is not a feature of all cognition, rather content-involving cognition is a special achievement. Intelligent beings capable of contentful thought will have participated in and mastered established socio-cultural practices — practices involving public representations that depend for their existence on a range of customs and institutions. Participating in such established socio-cultural practices is a necessary scaffold for the emergence of content-involving forms of cognition (see Hutto and Satne 2015).

What happens to basic minds in this process? Do basic minds become completely transformed when they are capable of dealing with contents? Do they become altogether different kinds of minds? Do they suddenly upgrade into a better class of mind, once they become capable of contentful cognition? Not on the REC view. Although superficially, talk of basic and non-basic, contentful minds may suggest it, REC denies that there are really two distinct and separate kinds of minds in operation here.

By REC's lights cognition is always dynamic, interactive, and engaging, but in some cases cognitive interactions are also content-involving. Even when cognition involves content and inferential processes the ultimate character of cognition remains enactive and dynamic. In other words, cognition can be content-involving without becoming content-based. Compare: Some conversations involve coffee, some don't. The former do not become coffee-based even if the introduction of coffee into the conversational mix makes a difference and has transformative effects on the way such conversations unfold. As it is with coffee, so it is with content.[4]

Introducing content into the cognitive mix does not change the fundamental nature or ultimate basis of cognition. Mastering certain practices can, quite dramatically, augment and add to our cognitive capacities. But though such mastery can add properties to cognition, opening up new possibilities for it, it does not—*pace* McDowell (1994)—completely overwrite the properties of our basic minds.

A major advantage of REC's commitment to content is that it provides a means of addressing the familiar "scaling-up" objection now regularly laid at its door. The scaling-up objection holds that basic cognition has important limits and that contents are needed to explain the many kinds of intellectual feats that lie beyond its reach. Yet in endorsing a duplex account of mind, REC can clearly allow that content-involving cognition is needed for some kinds of cognitive tasks—namely, any that require content, such as thinking true thoughts about boats on lakes.

Fundamentally, by REC's lights, basic cognition is a matter of sensitively and selectively responding to information, but it does not involve picking up and processing information or the formation of representational contents.[5] REC's account of basic cognition is thus given in terms of active, informationally sensitive, world-involving engagements, where a creature's current tendencies toward active engagement are shaped by its ontogenetic and phylogenetic history.

Basic minds target, but do not contentfully represent, specific objects and states of affairs. To explicate how this could be so Hutto and Satne (2015) introduced the notion of target-focused but contentless Ur-intentionality. In the next two chapters we elaborate and further motivate acceptance of that idea and show how it provides a sound basis for an account of the natural origins of truly contentful cognition.

5   Ur-Intentionality: What's It All About?

What is Ur-intentionality? It can't just be selective responsiveness to stimuli. Otherwise, we'll have to credit Eric Kandel's sea slugs with intentionality.
—Alex Rosenberg, "The Genealogy of Content or the Future of an Illusion"

Getting to the Bottom of Intentionality

Several chapters ago we announced that REC seeks to leave behind the claim that basic minds are contentful, while nonetheless holding on to the claim that they exhibit a kind of basic intentionality. This is easier said than done. On the face of it, any attempt to develop a tenable nonrepresentational alternative to cognitivism while also holding on to the idea that cognition is intentional seems plagued by internal conflict.[1] Or rather, as this section aims to show, such conflict arises as long as one endorses standard versions of what Roy (2015) dubs neo-Brentanism. What's neo-Brentanism?

Let's start with what Franz Brentano famously says about intentionality:

> Every mental phenomenon is characterized by ... the intentional (or mental) inexistence of an object, and what we might call, though not wholly unambiguously, reference to a content, direction toward an object (which is not to be understood here as meaning a thing), or immanent objectivity. Every mental phenomena includes something as object within itself, although they do not all do so in the same way. In presentation something is presented, in judgement something is affirmed or denied, in love loved, in hate hated, in desire desired and so on. (Brentano [1874] 2009, 124; Eng 88)

Brentano not only lastingly shaped philosophical thinking about which properties might define intentionality, he is also known for promoting two other ideas, that (1) intentionality is the one and only mark of the mental and (2) intentionality is not reducible to or explicable in terms of any natural phenomena. Adherents of neo-Brentanism happily stick with Brentano's basic characterization of intentionality, more or less, but they recast, modernize, and importantly soften his two claims above about its scope and status. Thus the new Brentano, cast in contemporary garb, tells us that "intentionality is one of the marks of the mental, as well as a naturalizable mark" (Roy 2015, 92).

Certainly, REC applauds both adjustments. But that is not where the apparent problem lies. The intolerable tension in endorsing nonrepresentationalism, on the one hand, and neo-Brentanism, on the other, lies in the fact, as Roy's analysis reveals, that to understand intentionality in standard neo-Brentanian terms is to think of it as having a representational essence. Unless that requirement is revised it would spell doom for the REC program. The remainder of this section aims to establish that it is conceptually possible to disentangle ideas that are unhelpfully run together in Brentano's classical formulation of the various properties of intentionality—a linking of ideas that makes representationalism seem inevitable. Once these ideas are disentangled, it is possible to think of the most primitive form of intentionality, which we call Ur-intentionality, in noncontentful, nonrepresentational ways while still making room for the possibility that the most basic kind of intentionality exhibits the trademark property of the intentional—that of being an attitude directed toward an object.

The first step toward that end is to note that Brentano's understanding of intentionality, which aims to explicate the doctrine of intentional inexistence, is he himself remarks, "not wholly unambiguous."[2] That verdict is unsurprising since Brentano's characterization of intentionality is complex, involving more than one proposed property. For example, he speaks not only of intentional phenomena being directed at objects and as referring to contents, but also of intentional objects being contained within mental phenomena, of their exhibiting an immanent objectivity.[3]

There are different ways of understanding these ideas in today's context. Yet those who look to Brentano for guidance on the nature of intentionality typically end up making an appeal to a notion of representation or content modeled *directly* on the kind of semantic content (whether truth conditional or referential) – content of the sort that is associated with propositional attitudes. It is by this route that many of today's philosophers come to endorse a content-based account of intentionality.

Flanagan (1991) supplies us with a neat reminder of today's received view of intentionality:

The concept of intentionality is a medieval notion with philosophical roots in Aristotle and etymological roots in the Latin verb *intendo*, meaning "to aim at" or "point toward." The concept of intentionality was resurrected by and clarified by ... Franz Brentano. ... Brentano distinguished between mental acts and mental contents. My belief that today is Monday has two components. There is my act of believing and there is the content of my belief, namely, that today is Monday. ... Beliefs are not alone in having meaningful intentional content. ... Language wears this fact on its sleeve. We say that people desire that [...], hope that [...], expect that [...], perceive that [...], and so on, where whatever fills the blank is the intentional content of the mental act. *Intentionality refers to the widespread fact that mental acts have meaningful content.* ... The fact that we are capable of having beliefs, desires, or opinions about non-existing things secures the thesis that *the contents of mental states are mental representations*, not the things themselves—since in the case of unicorns, ghosts, devils, and our plans for the future there simply are not real things to be the contents of our mental states! (p. 28, second and third emphases added)

How did this content-involving understanding of the essential nature of intentionality become the received view? Searle (1983) had a lot to do with it. Searle (1983, 1) defines the quarry in his landmark investigation into intentionality as "that property of many mental states and events by which they are directed at or about or of objects and states of affairs in the world." He pledges allegiance to a venerable tradition in taking "intentionality" as referring to the properties of mental "directedness" or "aboutness"—notions that he regards as equivalent. Searle (1983, 1) is motivated by his recognition that the term *intentionality* "is misleading and the tradition something of a mess."

Thus rather than relying on that tradition alone, he tries to isolate the phenomena of interest by using the following procedure: "If I tell you I have a belief or desire, it always makes sense for you to ask 'What is it exactly that you believe?' ... My beliefs and desires must always be about something" (p. 1).

The use of this procedure is quite telling. Naturally enough, if one focuses attention only on the kind of intentionality exhibited by beliefs and desires with articulable content it is easy to become convinced that intentionality must be, always and essentially, bound up with content—namely, that contents are both what one thinks about and that having mental contents is what determines or makes possible directedness or aboutness.

Muller (2014, 172) dubs the sort of approach Searle adopts a top-down approach to intentionality: it is one that starts "the story of intentionality at the end instead of the beginning." Muller adds that "philosophers of mind have arguably done the study of intentionality a disservice by focusing first and foremost on fully cognitive states such as beliefs and desires whose contents are both directed and individuated in terms of propositions containing what grammarians call complete thoughts" (p. 171).

The problem with the top-down strategy is that it dictates a limited set of answers to the questions about what the essential characteristics of intentionality might possibly be. This is beautifully exemplified in Searle's case. He gives what for many is the textbook answer to the question of the essence of intentionality, namely that "Intentional states represent objects and states of affairs in the same sense of 'represent' that speech acts represent objects and states of affairs (even though … speech acts have a derived form of Intentionality and … Intentional states … have an intrinsic form of Intentionality" (Searle 1983, 5).

In making these claims Searle (1983) is utterly explicit about the fact that he is directly modeling the intentional properties of mental attitudes on the properties of contentful linguistic utterances.[4] In doing this he appeals to our apparently clear intuitions about "how statements represent their truth conditions, about how promises represent their fulfillment conditions, about how orders represent the conditions of their obedience, and about how in the utterance of a referring expression the speaker refers to an object" (p. 5).

Recapping the main steps in his reasoning process, it is clear that by following Searle's top-down route, one will be led to think that intentionality has content necessarily—content of the same kind that is expressed by linguistic utterances. One can think this is so, even though the most basic form of so-called original intentionality is thought to have representational content that is mental, not linguistic. Indeed, the mental representational content of original intentional states cannot be identified with linguistic content if we are to have any prospect of explaining, as a familiar story would have it, how linguistic utterances come to have contents by borrowing them from purely mental sources. What is important to note is that the top-down strategy that has led to this familiar set of ideas has thoroughly and successfully captured the philosophical imagination, fueling the received view and keeping alive the belief that mental content of the sort that can be expressed in linguistic utterances is utterly essential for there to be any kind of intentionality, however basic.

The top-down strategy has resulted in a very narrow vision of intentionality becoming deeply entrenched in analytic philosophy. Today the thesis that intentionality is essentially content-involving is not simply treated as a possible way of understanding the phenomena. Nor is it generally noticed that it picks up on only one strand of Brentano's attempted explication of the notion. Rather, analytic philosophers have enshrined that assumption, maintaining that it captures the very essence of intentionality. This idea has become so ingrained that nowadays it is usual to define intentionality as necessarily representational. Thus the *Stanford Encyclopedia* entry on the topic tells us, with authority, that "intentionality is the power of minds *to be about, to represent, or to stand for*, things, properties and states of affairs" (Jacob 2014). Given these developments, it is hardly surprising that for many philosophers it is not just an unassailable truth, but utterly "uncontroversial that an intentional state has representational content: it is about something" (Muller 2014, 176).

The unshakable certainty that analytic philosophy now has a firm grip on the true nature of intentionality is ironic if we consider that "the notion of aboutness is itself mostly left unanalyzed, when not explicitly declared undefinable" (Roy 2015, 94).[5] Indeed, the fact is that in the contemporary literature, intentionality is "rather poorly defined [by appeal to] ... the intentional idiom itself. This is a problematic situation in many respects that cannot but reflect a theoretical embarrassment" (Roy 2015, 94).

Philosophers of a more naturalistic bent who are not impressed by the idea that we will discover the essence of phenomena through conceptual analysis that appeals to a priori intuitions are more open-minded about the diverse forms intentionality might take and what features these forms might have. Muller (2014, 169) states the case well, capturing what drives REC's attempt to recognize and understand Ur-intentionality:

For those of us interested in an account of intentionality that is both informed by evolutionary biology and consistent with the tenets of naturalism, states that are relatively basic, developmentally speaking, are certainly a better place to start than with states at the other end of the developmental scale. ... We must look for bottom-up approaches that seek to explain how full-blown cognitive and language dependent capacities might have been built-up from simpler, less abstract aspects of our developmental repertoire.

Summing up, Muller captures the mood of many working in the field today: "From a naturalistic perspective, one thing that intentionality is going to be is *something that is built up from simpler structures and capacities*, as opposed to something that appeared fully formed independent of what came before" (p. 169, emphasis added). With this in mind, what is needed is a more "nuanced understanding of intentionality" (p. 170). Hence, against the backdrop of these sorts of considerations, naturalistically minded philosophers have grown "skeptical of the idea that all representations have some essence in common" (p. 176).[6]

A familiar move in the literature is to dilute one's understanding of the kind of content and satisfaction conditions required for more basic forms of intentionality, such that the relevant norms are of a different and weaker sort than those to do with truth and falsity. This approach to understanding basic forms of content has become quite popular.

For example, Crane (2009) tries to articulate just such a nonpropositional account of content. He rejects the idea that perception is a propositional attitude yet without surrendering the idea that perceptual states possess representational content. Instead, he claims perceptual states have accuracy and correctness conditions that are not any kind of truth conditions. A guiding thought that motivates Crane's proposal is that perceivings can be more or less accurate or correct, as opposed to being simply true or false. Thus he holds that perceivings have a kind of representational content in that such states of mind present the world as being a certain way. Still, he denies that such content is of the truth conditional sort.[7] Perception, on this view, has a kind of content that is more primitive and more basic than that of propositional attitudes such as beliefs.

Nowadays, it is typical for contemporary theorists to question the idea that content is necessarily truth conditional. In this vein, we are told that "it isn't apparent that an intentional state, event, or object about something other than a state of affairs should be evaluated in terms of truth/falsity" (Gunther 2003, 5). Still, even for those prepared to go this far it is generally accepted that "any state (event, experience, and so forth) with content, is … governed by semantic normativity. For whether its content is conceptual or non-conceptual, propositional or not, an intentional state presents the world as being a certain way; and intrinsic to this presentation, to its content, is a set of (semantic) conditions under which it does this correctly, truthfully, satisfactorily, appropriately, skillfully, and so on" (Gunther 2003, 5–6). Accordingly, for those who hold this sort of view it is still the case that "semantic normativity is the mark of intentionality" (Gunther 2003, 6).[8]

A more radical view is possible; we can surrender the idea that basic forms of intentionality need involve correctness or satisfaction conditions of any kind. For example, in describing the kind of intentionality that characterizes absorbed coping, Dreyfus (2001, 148) holds that "the agent's body is led to move so as to reduce a sense of deviation from a satisfactory gestalt without the agent knowing what that satisfactory gestalt will be like in advance of achieving it." Likewise, drawing on Walter Freeman's work, Dreyfus tells the same story about the brain's contribution to absorbed coping, emphasizing that when understood as part of a larger dynamical system, a movement can achieve "satisfaction without the brain in any way representing the movement's success conditions" (p. 151).[9]

This is more or less the REC view: in basic kinds of cognition an organism's skillful engagements with the world are best understood in embodied, enactive, and nonrepresentational ways. Dreyfus (2002b) is clear that absorbed coping is more than an account of bodily activity that reduces to mere habitual reflexivity. The spontaneous responsiveness it exhibits is not to be understood in terms of blind reflexes.[10]

Despite agreeing substantially with REC on this issue, Dreyfus speaks as if such intentionality is contentful in some sense. Thus he claims that "even the most 'automatic' response to the solicitation of the situation must have content" (Dreyfus 2002b, 421). Explicating this view, he makes it clear that he supports the idea that "there are inner states of the active body that have intentional content but are not representational" (p. 414, original emphasis).

Despite this terminological difference, in abandoning the idea that all forms of intentionality require correctness conditions, Dreyfus and REC are both attempting to provide an understanding of basic world-relating attitudes in terms of a "nonrepresentational form of activity ... [that] is a more basic kind of intentionality" (Dreyfus 2002a, 377).

It is likely that what Dreyfus means by intentional content here is simply whatever object a given intentional attitude targets or is directed at. That would be an utterly anodyne use of the notion of content, and one that even RECers could embrace. However, since most analytic philosophers assume that content entails correctness conditions, to introduce talk of intentional content at this crucial juncture is likely to breed only confusion.[11] Hence REC recommends the keeping to the vocabulary of contentless intentionality rather than nonrepresentational intentional content.

Muller (2014, 179) reaches a similar conclusion at the end of his quest to try to explicate intentionality in terms of "simpler, less cognitive representations." He goes further than most, entertaining the possibility that "there is intentional content to which the sense-reference distinction does not apply" (p. 178). For him, that this might be so entails that there might be "representations to which the sense-reference distinction is applicable and those representations to which it is not" (p. 177). Weighing the considerations that push him toward this idea, he admits, "I don't know whether the suggestion that semantic content at this simple level is such that the sense-reference distinction does not apply will eventually pan out, but it merits further exploration" (p. 179).

As with Dreyfus's proposal, here again, we see only a nominal difference between Muller's suggestion about the properties of the simplest kind of representations and what REC proposes about the most basic form of contentless intentionality. We don't quite know what to make of the idea of representations that have semantic content but that lack both sense and reference, but—if worked through—we suspect this would equate with REC's contentless understanding of Ur-intentionality. It should be clear enough, despite Muller's representational gloss, that to come this far—to let go of the idea of a sense-reference distinction while retaining the idea of some kind of intentional directedness—is actually to go the REC way. Certainly, it is to abandon the idea that the most basic forms of intentionality must be contentful or representational in the sense of having sense, reference, or correctness conditions.

At last, we are finally in a position to address the puzzle described in the opening lines of this chapter. It was noted there that to accept the idea that intentionality entails the existence of any kind of contentful representation while endorsing antirepresentationalism is surely to embed contradiction in the heart of one's theorizing. The only way out for radicals is to try to devise a genuinely non-Brentanian or nonstandard neo-Brentanian understanding of intentionality (Roy 2015, 93).

What is needed is "a radically anti-representationalist kind of intentionalism" (Roy 2015, 117). That is precisely what REC aims to supply in proposing that Ur-intentionality is contentless—in proposing that the most primitive kind of intentionality is one to which the sense-reference distinction does not apply. Surely, if tenable, the notion of Ur-intentionality fits the theoretical bill. It is not surprising that in making this proposal, REC has been recognized as offering "the closest thing to an authentically non-representationalist intentionalism" (Roy 2015, 121).[12] The big question now remaining is: Is the notion of Ur-intentionality theoretically and naturalistically tenable?

Ur-Intentionality: The Natural Explanation

How can we understand the sort of world-directedness exhibited by contentless Ur-intentionality in naturalistic terms? The short answer is that this can be done by RECtifying teleosemantics—by stripping it of its problematic semantic ambitions and putting its basic apparatus to new and different theoretical use. Those adjustments will enable us to understand contentless forms of intentionality in naturalistic terms. So modified teleosemantics provides all that is needed for making sense of the special kind of target-focused, biologically based normativity exhibited by basic cognition. In short, appeal to a modified teleosemantics is REC's way of "explaining and understanding a contentless mind" (Shapiro 2014a, 218).

Classic teleosemantic theories seek to naturalize representational contents by appeal to biological function. At their core, such theories give pride of place to the purpose of device traits and responses, rather than focusing on dispositions and causes. Construed in semantic terms, such theories conceive of mental representations as inner states that have the biological function of enabling organisms to keep track of specific worldly items. Proponents of traditional teleosemantic theories understand the content of mental representations in truth conditional terms— thus a mental representation is true if things are the way an inner state represents them as being, and it is false in all other cases (Millikan 1984, 2005; Papineau 1987; McGinn 1989).

Importantly, not just any item with a biological function that systematically relates an organism to aspects of its environment qualifies as a mental representation according to such theories. Three conditions need to be jointly satisfied in order for this to be the case:

1. The item or response must be relationally adapted to some specific feature of the world.

2. The relation in question must be characterizable by a mapping or correspondence rule.

3. The item or response must have the proper function of guiding a cooperating consumer device in the performance of that consumer device's proper function with respect to tracking or acting on the world feature targeted in (1).[13]

Devices, traits, and states come to have their proper or biological functions through a history of selection by consequences. To understand the biological function of some item therefore requires giving explanations that focus not on an item's current dispositions but how it historically performed some function "on those (perhaps rare) occasions when [that function] was properly performed" (Millikan 1993, 86).

Explanations in terms of proper functions are not statistical, dispositional, or mechanical. Their concern is always with what consumer devices are supposed to do, not what they are disposed to do. Recognizing this is utterly crucial when it comes to understanding teleosemantic theories of content. It is for this reason that a device, trait, or structure can have a proper function even if it regularly fails to perform it. This is why, to borrow a timeworn example, the proper function of sperm cannot be understood only, or even primarily, by appeal to the actual behavior or dispositions of sperm.

An inner state has the proper function of representing some item if it is used by the organism to indicate or track the presence of that item—if it is "supposed to" do such work for the organism. The normativity implied by the "supposed to" operator is ultimately to be cashed out naturalistically by appeal to the past effects of species-wide selectionist pressures and individual learning histories. In sum, to understand an item's proper function requires explaining why it was nonaccidentally selected for because of what it did for the organism or organisms in question.

Teleosemantics is, far and away, the favorite strategy for naturalizing content: it is widely regarded as the "most promising" (Ritchie 2008, 161; Cash 2008, 104). Some of its fans even go so far as to say that teleosemantics is simply "inevitable," or that it is the only possible way of naturalizing content. For example, Rosenberg (2013, 3) insists that naturalism's "best resource, perhaps its only resource, for solving the basic problem of intentionality certainly seems to be Darwin's theory of natural selection."[14]

Despite the enthusiasm for teleosemantic theories and the common belief that they are our best hope for a naturalized theory of content, there are well-known, serious problems at the heart of such theories; recognizing this has led many of their most ardent fans to despair. For example, despite insisting that Darwinism is the only way we can hope to naturalize content, Rosenberg (2014b) also acknowledges that teleosemantics can't solve what Fodor (1990) calls the disjunction problem. From this he concludes: "So much the worse for original intentionality. ... If Darwinism about the brain can't give us unique propositional content, then there is none" (Rosenberg 2014b, 26).

We return to this point shortly, proposing that even if this damning assessment proves true, something of great importance can be salvaged from a modified teleosemantics, one that lowers its sights and aims for the lesser prize of understanding only Ur-intentionality rather than trying to explain mental content in semantic terms per se. We dub the modified theory teleosemiotics.

Before looking more closely at what motivates this approach, it is important to be reminded of the kinds of explanations of intelligent behavior that teleosemantics or any modified version of it can possibly offer. For whether such biologically based theories are used to naturalize mental content or only Ur-intentionality, understood from a teleological perspective, cognitive properties are only capable of doing limited sorts of explanatory work in the sciences of the mind. More precisely, if intentional properties are understood through the lens of teleological theories, then they are logically unable to do certain kinds of explanatory work *in the way* most cognitive scientists assume such work needs to be done.

Consider Spaulding's (2011) argument that aims to establish that at least some basic forms of nonverbal intelligent behavior cannot be adequately explained without assuming the existence of contentful representations. In attempting to make this case, Spaulding revisits Millikan's classic example of honeybees and their remarkable navigation abilities.

According to Spaulding, what best explains the sophisticated navigational feats of bees is that they "represent and protologically reason about the location of the hive" (Spaulding 2011, 156; see also Carruthers 2009, 98). In a bid to lend support to her conviction, Spaulding (2011) reports that she has "difficulty seeing" how anything other than a system of internal, well-structured representations with denotational and truth-evaluable contents could do the relevant explanatory work.

There is no doubt that teleological theories can explain why honeybees are set up so as to respond flexibly to the dances of their compatriots and thus coordinate their efforts to locate nectar. What such theories cannot possibly do, however, is explain how mental contents make a causal difference to the local processes responsible for such intelligent and flexible responding.

As noted above, teleological theories understand the intentional by appeal to proper functions, and when it comes to understanding proper functions, history sets the standard and supplies the norms. It is history that determines whether a current attitude represents correctly or incorrectly or what it is directed at. Consequently, understood via teleological theories, intentional properties are debarred from playing any kind of causal role in the synchronic production of intelligent behavior.

It follows that, according to teleological theories, intentional properties are patrician and not plebian in character. In other words, the intentional never gets its hands dirty. The intentional, if understood in terms of biological functions, does no causal, mechanistic work.[15] This is why teleological theories provide explanations in terms of ultimate as opposed to proximate explanations – in terms of structuring as opposed to triggering causes (Dretske 1988, chaps. 1, 2).

Intentional explanations that cite proper functions look to the historical conditions under which the relevant biological devices were selected for; such explanations only indirectly answer questions about how such devices are likely to operate in the here and now.

Consequently, to adopt a teleological take on intentionality is to forgo the idea that intentional properties can possibly feature in proximate mechanistic or causal explanations. For that reason, when understood through the lens of a teleological theory, the intentional properties cannot possibly do the sort of work that cognitivists such as Spaulding insist mental contents must do in driving intelligent behavior. In particular, intentional properties construed along teleological lines cannot answer "how-questions about cognitive capacities, [as opposed to] ... why-questions about particular behaviors or actions" (Gładziejewski 2015, 66).

This salient reminder should concern cognitivists, like Shea (2013), who appeal to teleological theories and yet also hold that the reason we need to posit mental content is precisely to answer very particular kinds of how-questions.[16] In describing the important explanatory need to say how systems are connected to aspects of their worlds, Shea (2013) makes it clear that in his view,

> What adverting to content does achieve, however, is to *show how the system connects with its environment*: with the real-world objects and properties with which it is interacting, and with the problem space in which it is embedded. The non-semantic description of the system's

internal organisation is true of the system irrespective of its external environment. Content ascriptions help explain *how it interacts with that environment*. (p. 498, emphasis added)

A teleological theory of the intentional cannot possibly address the connection question that Shea asks. At most, a theory of that kind can only explain why a system connects to and targets certain features of its environment; it is in no position to say or explain *how* systems connect to their environments. Given that teleological theories can't answer how-questions, it follows that they cannot explain how content might be supplied to the brain via the senses or how it might causally mediate and drive behavioral responses.

This point is made vivid by focusing on Gallistel's (1998) highly influential work on insect navigation. Gallistel invokes talk of representations in the explanations he offers of such behavior, but the only items doing actual load-bearing work in his explanations are systematic structure-preserving correspondences—correspondences that hold between certain features of the organisms and certain features of their environments. Although the bees' exploitation of those correspondences feature heavily in Gallistel's explanations of their navigational behavior, contentful representations make no appearance at the level of cognitive drivers of such behavior. In discussing this very case, Rescorla (2012b) makes clear that, "Explanatory power resides solely in the 'functioning isomorphism' between mind and world. There is no obvious reason why 'functioning isomorphism' must have truth conditional content. … The burden of proof lies with those who claim that functioning isomorphism suffices for truth conditions" (Rescorla 2012b, 96; see also Tonneau 2011/2012).

Taken together the above observations highlight the most serious problem facing teleological theories of content: The fact that thinking of content under the auspices of such theories debars mental content from featuring in explanations in the sciences of the mind or biology about the items and processes that proximately guide intelligent, flexible behavior or explain how systems connect to the world. Looked at from a purely biological point of view there appears exactly no reason to believe and no ground for supposing that the connections forged between organisms and features of their worlds are best understood as instantiating semantic relations involving reference or truth. Indeed, from a purely biological perspective that assumption appears to be explanatorily superfluous and extravagant—it does no work.

Put simply, there is no obvious reason why basic biologically forged mind-world connections ought to be characterized in semantic terms. Why should simply having a biohistory—one that determines which worldly offerings a creature is meant to target—be thought to entail the existence of any kind of semantic properties? There is no apparent scientific reason for supposing that the connections between an organism and its world that are established by Mother Nature are best characterized in semantic terms, such as reference or truth.

This conclusion will not come as a surprise to anyone who knows what originally motivated Millikan to propose her teleosemantic theory of content. That is because her theory was devised to satisfy a particular philosophical agenda. Millikan's prime purpose in creating teleosemantics was to answer a very specific philosophical need—a need quite distinct from what is required to deal with the explanatory projects of cognitive science or biology.

The driving ambition that led to the development of Millikan's teleosemantics was to provide an updated version of Wittgenstein's so-called picture theory, in order to complete Sellars's project of connecting the big ideas of Wittgenstein's early and later thinking about mind and language. For that purpose, Millikan (2005, 77) explicitly "pursued picturing themes." It was as part of this philosophical effort that she turned to natural selection to try to explain how there can be fully objective, natural correctness conditions that are disposition-transcendent.[17]

The aspiration to give a naturalized account of rule following, and not an ambition to be true to the needs of biology, is what drives her to say, using that favorite philosophical example, that "bee dances have truth conditions. The rules by which they are designed to correspond to nectar locations are semantic rules" (Millikan 2005, 97–98). The hope that teleosemantics lives by is that if we can find semantic rule following in the natural world at a very basic level, then it will be possible to explain the semantics of language without further ado. Millikan (2005) explicates this crucial link as follows:

The intentionality of language is exactly parallel to the intentionality of bee dances. ... Language forms ... have ... a function or series of functions ... [such that] ... if ... the form will guide the hearer so that its stabilizing function is performed *only when there is a correspondence by a given rule or function between the form and some structure in the world*, then the form is intentional. It has a truth condition. (p. 98, original emphasis)

The point is that, for Millikan, a teleosemantic theory of content is meant to supply the means of connecting the picture theory of Wittgenstein's *Tractatus Logico Philosophicus* (1922) with Wittgenstein's later account of language, as presented in his *Philosophical Investigations* (see Millikan 2005, 77). Proposing that these views are compatible "in basic measure," Millikan hoped to use her biological theory of content to supply a robust, disposition-transcendent explanation of the normative features of factual representation that she found to be missing in the work of the later Wittgenstein.

Crucially, Millikan's whole point in proposing a teleosemantic theory of content is thus to supply a pivotal but missing naturalized account of the norms needed to explain the semantics of language. According to her, a naturalized account of such norms is necessary to avoid a crippling shortcoming of an unsupplemented *Investigations* perspective on meaning. For, reading *the Philosophical Investigations* through a Kripkensteinian lens, Millikan (2005, 84) maintains—quite incorrectly, as it happens—that on Wittgenstein's later view of meaning "the criterion for having followed a rule can only be public agreement."

To avoid the admittedly disastrous idea that truth conditions are established by public convention, Millikan sought to establish that Mother Nature is the independent source and arbiter of standards of truth and correctness. It was this philosophical need, and not any observations stemming from biology, that drove her to claim that in living systems that exhibit intentional directedness, even at the "simple level a stringent criterion for rule-following is in effect" (Millikan 2005, 87).

The problem with Millikan's plan is that, if we stick only with what biology can supply, it has been shown that "teleosemantics can't individuate intentional content. No amount of environmental appropriateness of a neural state or its effects is fine-grained enough to give unique propositional content to the neural state" (Rosenberg 2014, 26; for more details see also Hutto and Myin 2013, chap. 4).

The bad news is that if we stick solely and parsimoniously only to what biology can really provide, it is clear that "Millikan has not provided an adequate theory of content. Millikan's technical apparatus *does define a relation that can hold between a system's mental state and properties sometimes instantiated in the environment*. But ... the relation so defined is not 'has as its content that'" (Pietroski 1992, 268, emphasis added).

The good news, as the emphasized part of the above quotation highlights, is that even if this verdict is accepted all is not lost—appealing to biological functions to understand intentionality can still explicate "an important kind of natural involvement relation ... [even if that relation is] not ... representation or anything close to it" (Godfrey-Smith 2006, 60). In sum, we agree with Muller's (2014, 158) assessment of the situation: "It isn't the intentionality or aboutness of the mental that is the problem, it is thinking of that content as specifically propositional that is the problem."

Objects and Objections

We have now come full circle. Can REC, through teleosemiotics, provide what is needed to explicate Ur-intentionality in a suitably neo-Brentanian way that does not conflict with its antirepresentationalism?

Roy (2015, 123) says of REC that it "clearly points in the right direction, since, as a theory of representation, teleofunctionalism is a theory of objectivation." What is objectivation? As Roy (2015) uses the term, it picks up on one of the strands in Brentano's analysis of intentionality—it denotes "the very fact that the relation is one to something as an object ... in the sense of a relation in which the relatum is made into an object by contrast to relations in which it is a mere thing" (p. 95).

How should we understand the way of being related to a worldly offering such that it is an object for the organism or system and not a mere thing? The timeless philosophical example—the case of the humble frog and his crude feeding habits—provides a convenient way of explicating REC's nonrepresentational take on objectivation. Frogs are inclined to lash out with their tongues when presented with small, dark stimuli that move in ways that are sufficiently like the movements of flies, their traditional prey. Thus frogs reliably respond to a range of different things that exhibit this signature behavior—and that list includes many things that are not flies, such as BBs or shadows moving in the right way.

Why, given this behavioral profile, should we think the frog is targeting flies and not all the other items on the long list of things it is disposed to chase? What entitles us to think that flies, and not all those other things as well, are the intentional objects of the frog's responsiveness? That hypothesis is justified by the fact that the frog's perception-action routines were forged through a long process of selection by consequences in order to get flies, and not anything else, into its belly.

Hence, the biological function of the frog's tongue-snapping behavior is not disjunctive at all; it is directed at flies and flies alone. This is because ingesting the flies, as a matter of fact, met the needs of this type of frog's ancestors. Consequently, "targeting flies" alone enters into a full explanation of why the frog's perception-action responsiveness was first forged and then proliferated (Millikan 1991, 156).[18]

It is these biological facts that fix what the frog is directed at and explain why it is so directed, and hence why it is connected to this worldly offering and not some other. It is also explains why—in a limited, nonsemantic sense—should it chase shadows or BBs it responses would be misaligned and inappropriate.

Crucially, REC tweaks the standard teleosemantic story by dropping any commitment to understanding these sorts of basic intentional relations as involving semantic properties such as reference and truth. This adjustment provides REC with the resources to answer to Roy's pressing question: What does the directedness of behavior consist of "if it cannot be 'inherently say[ing] anything about how things stand in the world?'" (Roy 2015, 123). Also, assuming the proposed RECtification of the notion of sense making in chapter 4 is tenable, it is possible to explain what it means for organisms to respond to particular worldly offerings as objects of significance rather than interacting with them neutrally as mere things (cf. Roy 2015, 123).

It should now be clear why REC's nonrepresentationalist account of Ur-intentionality and objectivation does not reduce to a crude story about "a form of behavior" (Roy 2015, 95). REC retains the idea from teleosemantics that intentional directedness has a normative dimension such that it does not reduce to mere behavior or dispositions.

It is because REC casts Ur-intentionality in normative terms that it does not equate basic intentional directedness "to a sort of property of natural attunement and thus loses its connection with ... objectivation" (Roy 2015, 123). The natural attunements between organisms and their environments in the past not only structure the profile of an organism's current tendencies for response, they normatively fix what is intentionally targeted, in complicated ways across multiple spatial and temporal scales.

Indeed, it is precisely because REC makes room for at least this much normativity that it differs from the eliminativist, strict naturalist approaches, as promoted for example by Alksnis (2015), Abramova and Villalobos (2015), and Rosenberg (2014a, 2014b, 2014c, 2015).

Roy (2015) overlooks the fact that REC's account of Ur-intentionality has this crucially important normative dimension when he complains that REC is silent on the following question: "In what does responding behaviorally to something as an object correspond exactly behaviorally speaking?" (Roy 2015, 123). It should be clear, in light of the above, why REC's silence on this score is a studied silence.

To ward off another possible confusion, it should be clear that the history of attunements that fix the norms in question is not always directly tied to long tracts of ancient evolutionary history. The initial connections to the world that Mother Nature forged across evolutionary time are not—at least for most animals, and certainly not for humans—inflexibly fixed and hardwired. New connections can be actively forged between organisms and aspects of their world through experience. An organism's tendencies adjust as it gravitates toward an optimal grip on situations through individual learning. Such adjustments establish new transient norms for perceiving and acting, along with new anchors for attention, through an ontogenetic process of selection by consequences.

Bruineberg and Rietveld (2014) describe how experience enables organisms to adjust and adapt their behavior as they tend toward patterns of responsiveness that offer more effective ways of coping with a multiplicity of possibilities for acting in particular circumstances. The interplay of itinerant dynamics at different time scales—the complex mix of slower- and quicker-evolving dynamics—is what explains how organisms can be "both robust and flexible" (p. 10).

The point can be illustrated by the example of experienced boxers who adjust their striking patterns relative to their distance from boxing bags. They attune to the possibilities for action afforded to them in such circumstances over time and through practice (Hristovski et al. 2006; Hristovski, Davids, and Araújo 2009). Citing work by Chow et al. (2011), Bruineberg and Rietveld (2014, 10) explain how boxers tend to settle on an "optimal metastable performance region where a varied and creative range of movement patterns occurred." Without knowing it, expert boxers naturally seek out this zone of optimal metastable distance because it "offers a wide range of action opportunities and the possibility to flexibly switch between them in line with what the dynamically changing environment demands or solicits" (Bruineberg and Rietveld 2014, 10). This tendency toward optimal metastable attunement to the dynamics of the environment is a prime example of how individual experience, learning and practice can continually adjust our naturally installed factory settings.

At this juncture, it is worth correcting a certain misreading of the REC view—one that has arisen from our tendency to focus on the classic but simple example of the intentional directedness of frogs when discussing Ur-intentionality. Based on its excessive attention to the frog case, Kiverstein and Rietveld (2015, 712) worry that REC's account of intentionality "misses the skill."[19] This is because the mechanism in the frog is set up in a hardwired way "to be triggered only by specific features in the environment" (p. 708).

On Kiverstein and Rietveld's (2015) reading REC assumes that core features of the frog case apply universally to all cases of basic cognition, without adjustment. Accordingly, as they see it, REC's account of basic cognition reduces, ad absurdum, to the view that even very complex organisms, such as ourselves, are only ever capable of responding to a single affordance at a time and only by means of slavishly fixed and inflexible routines. In a word, to accept this reading, REC's vision of basic cognition would be utterly cycloptic—we and other creatures would only be able to respond to possibilities for action, one at a time, from singular, fixed, and inflexible perspectives.

For the record, let us be clear. REC fully agrees with Kiverstein and Rietveld (2015, 712) in thinking that the "animal is always ready to respond to more than just the affordance that it is currently acting upon." Moreover, REC agrees that this responsive openness to environmental opportunities operates in multiple ways across multiple spatiotemporal scales.[20] This way of understanding skilled responsiveness is wholly congenial to and compatible with REC's take on Ur-intentionality.

Yet REC's take on the Ur-intentionality exhibited in basic cognition provokes another important question: If we cannot simply look to natural selection for answers about the origins of content, then where, if anywhere, do we find content occurring in the natural world, and how do we explain it?

REC's answer is that it is only when very special forms of sociocultural practices are mastered that the "question of truth" can arise. Thus RECers agree with Huw Price's (2013) un-Millikanesque answer to the question of where we find factual representation in nature. Price (2013) is interested in our capacity to get things right or wrong in a way that does not reduce to communal agreement. On Price's analysis, making judgments and thinking thoughts that can be true or false only becomes possible when "something plays a role that is answerable" to "the kind of 'in-game externality' provided by the norms of …

[a] game [in which] players bind themselves, in principle, to standards beyond themselves" (p. 37).

Telling the story of the natural origins of content in full would require saying how basic minds came to master the relevant public practices that made it possible to fix the right kinds of standards. As a prelude to telling that story in greater detail elsewhere, the next chapter aims to allay the concern that a naturalistic story along these lines is impossible—that it simply cannot be told.

6 Continuity: Kinks Not Breaks

The difference in mind between man and the higher animals, great as it is, certainly is one of degree and not of kind.
—Charles Darwin, *The Descent of Man*

Getting Radical about the Origins of Content

Enactivists of all sorts emphasize the role of active, embodied engagement over representation when it comes to understanding cognition. For Radical Enactivists about Cognition, RECers, this is not just a matter of emphasis: they advance a stronger claim, holding that (1) not all cognition is content-involving, certainly not its root forms (Hutto and Myin 2013). Even so, RECers are not content deniers; they do not embrace global eliminativism about content.

RECers hold that, if appearances don't deceive us, then (2) some thoughts and linguistic utterances are contentful. Indeed, RECers allow that such thoughts and utterances—canonically judgments about factual matters—are contentful in the full-blooded sense of exhibiting the familiar semantic properties of reference and truth.[1] That (2) is the case appears not only to be borne out by experience, it is necessary for explaining prominent features of at least some forms of cognition. What's more, RECers are naturalists, albeit of a relaxed sort. They hold that (3) it is possible, in principle, to explain the origins of content-involving cognition in a scientifically respectable, gapless way. RECers aim to do so by making special reference to the important role played by sociocultural scaffolding (Hutto and Myin 2013; Hutto and Satne 2015).

As a prelude to telling that story in detail, this chapter responds to charges that a REC-inspired program for explaining the natural origins of content—the NOC program, for short—is doomed to fail. It addresses the general concern that the NOC program is internally incoherent when seen in light of the HPC. Then it defuses a second complaint about the NOC program—that in drawing a sharp distinction between basic, contentless and content-involving kinds of cognition, REC is at odds with evolutionary continuity. Finally, after showing that there is no reason to rule out the possibility of telling the story of the natural origins of content in a gapless, REC-friendly manner, an initial sketch is provided that shows the basic contours of how such a story might be told.

REC's Fatal Dilemma?

It has been claimed that the trio of REC assumptions outlined above is incompatible (Alksnis 2015; Korbak 2015). If REC's critics are right, its central commitments do not form a coherent set. An apparent symptom of this incompatibility is RECers allegedly inconsistent application of the Hard Problem of Content, or HPC. RECers invoke the HPC in order to motivate adoption of a contentless view of basic minds. But RECers seemingly ignore the HPC's force, selectively, when allowing that some minds—the subset that have benefited from the right kind of sociocultural scaffolding—have a contentful character.

The attempt to use the HPC in this selected way, so the critics of REC insist, is confused. They hold that if there is a HPC then it must afflict *any and all* attempts to explain how content could be part of the natural order. REC's critics deem the HPC to be a universal acid—an acid that, once out of its bottle, cannot be contained in the way RECers have hoped to confine it when pursuing the NOC program.

To be consistent, RECers should hold either that content can be understood in a suitably anaemic way such that it makes an appearance wherever we find minds, or that the existence of content as a naturally occurring phenomenon should be denied across the board.[2] Among those who see the situation in this all-or-nothing manner, anyone hoping to explain how content might emerge through sociocultural scaffolding, as RECers do, faces not merely a difficult challenge but a fatal dilemma.

As Alksnis (2015, 674) formulates the problem: "The first horn of the dilemma is a rejection of the compatibility of content with naturalism; the second is the rejection of content in order to preserve naturalism." According to these assessments, as long as REC sticks with explanatory naturalism, it will find it impossible, in principle, to explain how creatures that begin life with only basic contentless minds could ever come to have minds of a content-involving sort.

Undeniably, RECers make much of insuperable difficulties that the HPC poses for restrictive naturalists—namely, naturalists who limit themselves to *using only a narrow set of resources* when accounting for the place of contentful states of mind in the natural order. That is true. Yet there is no inconsistency in holding that the HPC is a hard or even impossible problem for some explanatory naturalists but not others.

How so? Simply put: different resources, different explanatory prospects. Thus the problem of content is only hard—impossibly hard—for naturalists who limit themselves by using overly narrow resources when trying to deal with it. RECers avoid the HPC by making appeal to a new set of expanded explanatory resources. In line with a neo-pragmatist tradition, RECers maintain that "the primary bearers of content are semantically articulated symbols, occurring in appropriate dynamic patterns" (Haugeland 1990, 412). The job is then to seek to explain how contentful states of mind actually come into being through a process of mastering special kinds of sociocultural practices (see Clapin 2002, 17–18, Haugeland 1998).

RECers assume that such explanations, supplied in a naturalistic register, are at least possible. If so it follows that the HPC does not apply universally to every variety of explanatory naturalism. Thus RECers think that whereas it is impossible to explain content using only the limited resources of restrictive naturalism, it is entirely possible to explain the origins of content, at least in principle, using the expanded resources afforded by a relaxed naturalism. This can be done in a way that does not presuppose the prior existence of content (Hutto and Satne 2015).

A relaxed naturalism is one that avails itself of the full range of scientifically respectable resources, drawing on the findings of a wide variety of sciences that include not just the hard sciences but also cognitive archeology, anthropology, developmental psychology, and so on.

Many explanatory naturalists are not relaxed but restrictive naturalists: they seek to naturalize content by using only the resources of the hard, natural sciences (causation, informational covariance, biological functionality) and nothing more. Such naturalists see no prospect of trying to explain content by appeal to sociocultural factors. There are two main arguments—one general, one specific—that motivate imposing such tough restrictions on any attempt to explain the natural origins of content. It is important for REC to disentangle and defuse both of these motivations. Let's take them in turn.

Strict naturalists are motivated by an uncompromising unification agenda. They insist that all good naturalists must subscribe to such an agenda. So if RECers are naturalists at all, they have no choice but to use only the restricted resources of the hard sciences. This is a general argument for restrictive naturalism.

Extreme naturalists of this kind hold that reductionist explanations are required of any phenomenon if it is to qualify as bona fide natural (Rosenberg 2015; Abramova and Villalobos 2015). They maintain that "reality contains only the kinds of things that the hard sciences recognize" (Rosenberg 2014c, 32). They also insist that "natural science requires unification" (Rosenberg 2014a, 41). The combination of these views adds up to zero tolerance for any phenomenon or domain that won't reduce. For example, Rosenberg (2014c, 41) holds that "science can't accept interpretation as providing knowledge of human affairs if it can't at least in principle be absorbed into, perhaps even reduced to, neuroscience."

A serious concern about this austere program is that if naturalists are "too exclusive in what they count as science, naturalism loses its credibility" (Williamson 2014b, 30). An obvious criticism of the extreme naturalism agenda is that it imposes overly strong, extrascientific ideologically driven constraints on scientific inquiry—constraints that we have little or no reason to suppose will pay off in the end (see Horst 2007, 21).[3] For these reasons we do not take this line of argument to have any real force and will not discuss it further here.

The second, special argument for restricted naturalism needs more attention. It offers a prima facie more plausible rationale for adopting restrictive naturalism—namely for using only the sparse tools of the hard sciences when attempting to naturalize content. It is driven, not by general reductive commitments, but by the belief that such restrictions on naturalism's legitimate explanatory resources are necessary because of special features of this particular explanandum. In particular, it is assumed that content *must be* in place before any sociocultural practices appeared (or appear) on the scene. The latter assumption captures the thought that the existence of the relevant kinds of sociocultural practices depends on the logically prior existence of content in ways that make it impossible for the former to explain the genesis of the latter.

Thus, if there is content, then restrictive naturalists of this second stripe assume we have no choice but to accept that it originates in states of mind that necessarily exist quite independently of, and ontologically prior to, sociocultural practices. This line of argument was made prominent in Fodor 1975 and Searle 1983, and even today influential thinkers take it for granted that "external symbols acquire their meaning from meaningful thoughts—*how could it be otherwise?* ... [External] symbols [cannot] be meaningful independently of the thoughts they have been designed to express. So, if the Hard Problem spells doom for contentful thinking, it ought to spell doom as well for our abilities to understand and use language" (Shapiro 2014a, 217–218, emphasis added).

To assume that content can *only* derive from mental content rules out a priori the possibility of explaining the natural origins of content by appeal to the mastery of sociocultural practices. We are meant to be persuaded by the "how could it be otherwise," "there's no other way" addendum. But surely there is an alternative explanation staring us in the face. It is that one that REC promotes: namely, that contentful thoughts only become so when the use of external symbols is mastered.

The possibility of a workable naturalistic account of the sociocultural origins of content is apparently ruled out if: (1) participating in and mastering sociocultural practices requires cognition; (2) cognition entails intentionality; and (3) intentionality entails content (for a fuller discussion, see Hutto and Satne 2015).

Certainly, REC's proposal about the sociocultural origins of content would embed a crippling, essential tension if it is agreed that: "original intentionality is essentially a social institution" (Haugeland 1990, 414), where this entails that "not only specific types of thought but thought in general depends on language" (Barth 2011, 1). For then there would be "no room whatever for original intentionality in any animals, (asocial) robots, or even isolated (unsocialized) human beings" (Haugeland 1990, 414).

REC avoids this problem because it rejects a pivotal assumption –one made by nearly everyone in the naturalizing content game. Specifically, in line with the analysis of chapter 5, it denies that "to have intentionality is to have (semantic) content" (Haugeland 1990, 384).

There is no logical contradiction in REC's proposed sociocultural account of the natural origins of content so long as basic cognition exhibits only contentless intentionality. If that is allowed, and HPC only spells doom for other restrictive attempts to explain the origins of content without appeal to sociocultural practices, then there are no grounds, *pace* Shapiro, for ruling out REC-inspired attempts to explain how we come to understand and use contentful language.

To recap, there is no inconsistency in holding, as RECers do, that the HPC troubles only some explanatory naturalists and not others. Nor does Shapiro cite any reason—at least any that does not blatantly beg the question—for denying outright the possibility of naturalistically explaining how contentless minds might become content-involving through a process of sociocultural scaffolding.

Evolutionary Discontinuity?

Suppose it is allowed that a REC-motivated project of trying to explain the natural origins of content is not internally confused or simply impossible. Still it might be thought that REC is hopeless for a different reason. It might be thought that if REC's distinction between basic, noncontentful and nonbasic, contentful minds were admitted this would entail the existence of a deep discontinuity in nature (or at least in how we are to understand nature). Menary (2015a) explicates:

Radicals have a problem bridging the gap between basic cognitive processes and enculturated ones, since they think that meaning, or content, can only be present in a cognitive system when language and cultural scaffolding is present (Hutto and Myin 2013). That, of course, *doesn't sit well with evolutionary continuity* (p. 3n5, emphasis added).[4]

What exactly is the problem? Why should embracing the idea that there are both contentless and contentful forms of cognition entail evolutionary discontinuity? A little probing reveals that, according to REC's critics, the ultimate source of trouble lies with its assumption that "basic minds should not be characterised as representational but that language users take on some of the representational capacities of the language they use; let us call this *the saltationist view*. On the saltationist view representation is only added on as a consequence of using language, narrative, or possibly some other social resources" (Clowes and Mendonça 2016, 17, emphasis added).

REC is apparently committed to saltationism because it assumes that the arrival of content-involving minds depends on mastery of specific kinds of sociocultural practices—making content utterly unprecedented in nature.

In advancing this view about the sociocultural origins of content-involving cognition REC allegedly *"poses a chasm between representationally enhanced and more basic minds"* (Clowes and Mendonça 2016, 18, emphasis added). And, so the complaint goes, once such a chasm is introduced we can rule out any possibility of providing a naturalistic explanation of the emergence of content in terms of gradual, continuous change of the sort evolution apparently favors.

This sort of worry has been used, dialectically, as a reason for embracing one of two more monomorphic visions of cognition—those that insist on psychological continuity across the board. Insisting on the need for psychological continuity has motivated acceptance of one or another of the single-storey stories about cognition. REC is thus flanked on the right, by "content-everywhere" views of cognition, and on the left by "content-nowhere" views of cognition. The right holds that Cognition always and everywhere Involves Content, aka unrestricted CIC. The left seeks to eliminate content altogether in advancing, for example, the Really Radical Enactive, Embodied account of cognition, or RREC (Harvey 2015; van den Herik 2014; see also Myin and Hutto 2015). Opting for either of these accounts, but not REC, apparently offers special protection against the nasty consequence of evolutionary discontinuity.

A comparison proves instructive. Unlike REC, Sterelny (2015) maintains that, although standard-variant teleosemantics is limited in key respects, it nonetheless provides the right resources for thinking about the content of basic kinds of minds (Millikan 1984, 1993, 2004, 2005; Papineau 1987; McGinn 1989).[5] To assume that teleosemantics is essentially correct, if limited, is to assume that minds that occupy the lowest rung of mindedness are contentful, in a certain rudimentary way.

This assumption about basic minds fits with the CIC assumption that all minds are contentful to some degree. But this still leaves room for CICers to allow that human minds are contentful in uniquely special, impressive ways. Thus, for example, it is possible for Sterelny (2015) to recognize the remarkable differences between basic and distinctively human forms of cognition, while simultaneously holding that these differences should only be considered a matter of degree rather than of kind.

A popular solution is to posit a sliding scale of content that is linked with increasing flexibility of response. Drawing on consumer-based teleosemantics, for example, it has been argued that:

the very same physical signal (say a pattern of light) that carries no semantic information for a rock (no interests, no actions available), minimal semantic information for a tree ("grow this way, more sunlight to the left"), and moderate semantic information for a rat ("more trees, fewer predators, shelter there") can carry much more for us ("saplings left, old growth right—this must have been where the old farm's boundary was"). (Cao 2012, 59)[6]

The upshot is that if all minds are assumed to be contentful, à la CIC, then there is apparently no explanatory barrier to understanding how "complex agents evolved incrementally from simpler ones" (Sterelny 2015, 552). With the CIC assumption in place, for example, Sterelny is free to hold that once the origins of our scaffolding practices are explained, and "how they work, and what their effects are, we are done. There is no extra problem then of explaining intentional content" (p. 562).

On this view, whatever other differences there may be in their respective cognitive profiles, at rock bottom, there is an assumed "*psychological continuity* between human and nonhuman animals" (Bar-On 2013, 315). What is on offer here is an assurance of gapless evolutionary continuity based on the assumption of a fundamental psychological continuity. This psychological continuity putatively exists along the full spectrum of mindedness and is ultimately cashed out in unrestricted CIC terms. Of course, all of this is perfectly in tune with the letter of Darwin's own take on the issue, and what Penn, Holyoak, and Povinelli (2008, 109) identify as the dominant trend in comparative cognitive psychology, since both regard any relevant cognitive differences between simpler and more complex minds as a matter of degree and not of kind.

Apparently, going the CIC way on the psychological continuity question has the added advantage of skirting the Scaling Down Objection that Korbak (2015, 92) levels against REC.[7] According to the Scaling Down Objection, "Whatever it is that makes human linguistic practices give rise to content should also give rise to content in waggle-dancing honeybees, quorum-sensing in bacteria, and hormones in one's endocrine system" (Korbak 2015, 94). The driving assumption motivating the Scaling Down Objection appears to be that to the extent that content is deemed a naturally occurring psychological phenomenon at all, then for the sake of consistency, we must find it, at least in some modest form, everywhere in the domain of the psychological.

So again, what makes the unrestricted CIC view attractive, allegedly, is that unlike REC, it can tell a homogeneous tale about content that "does not need a sophisticated account of spontaneous generation of an elaborate discourse" Pattee (1985, 26). Following Pattee's logic, Korbak (2015, 93) holds that on the contrary, "Highly evolved languages and measuring devices are only very specialized and largely arbitrary realizations of much simpler and more universal functional principles by which we should define languages and measurements."

A pertinent question to ask about the CIC "content-everywhere" proposal is: Must psychological continuity be posited at all in order to avoid introducing gaps into the evolutionary story? Or, more simply: Does evolutionary continuity logically require psychological continuity? Before answering too quickly we should consider the objection that "the conviction that there *must be* some diachronic emergence story encourages proponents of continuity to over-interpret the mentality and communicative behaviors of existing animal species, and to underplay some of the evidently unique features of human thought and language" (Bar-On 2013, 296, emphasis added).

Concerns about overinterpretation invite a second question: Is positing CIC-style psychological continuity really the best way to account for the distinctive properties of human minds? Some decidedly think not; indeed they take it to be "one of the most important challenges confronting cognitive scientists of all stripes ... to explain how the manifest functional discontinuity between extant human and nonhuman minds could have evolved in a biologically plausible manner" (Penn, Holyoak, and Povinelli 2008, 110).

Finally, in assessing all of this, as Crane (2014) cautions, we must carefully specify what we mean by "fundamental," "degree," and "kind" when speaking of fundamental differences in kind, not degree. On this, he writes that "if a fundamental difference is just an important difference, or a significant difference, or a scientifically or philosophically interesting difference, Darwin's claim is surely not true" (p. 140).

With this list of concerns in mind, it is natural to ask if there are scientifically tenable alternatives to the psychological continuity demand, or at least the standard unrestricted CIC way of satisfying it, that might account for the important features of human minds that apparently set them apart. As noted, REC assumes that there are good independent reasons for supposing that when it comes to minds it is a mistake to think of them as possessing content "all the way down."

In particular, REC holds, *pace* Sterelny, that teleosemantics fails to deliver its promised theory of content and this is precisely because the violation of mere biological norms is never a matter of misrepresenting how things stand with the world. In this respect, REC agrees with Burge (2010, 301)—and many others—in maintaining that there is "a root mismatch between representational error and failure of biological function" (also see Pietroski 1992; Putnam 1992; Haugeland 1998; Price 2013). It doubts, *pace* Millikan (2005), that appeals to biological function can solve the rule-following problem.

As noted at the end of the preceding chapter, we follow Price (2013) in emphasizing a very special sort of practice that opens up the possibility of getting things right or wrong in ways that do not reduce to purely biological norms or even norms of communal agreement or conformity. The possibility of contentful error requires being a participant in a practice in which the question of truth can arise for what one thinks or says.

On the REC view, the unique sensitivity to very particular kinds of norms needed for playing the latter type of game is a necessary requirement for the existence of contentful ways of thinking and talking. REC holds that the development of such intersubjective practices and sensitivity to the relevant norms comes with the mastery of the use of public symbol systems.

As it happens, this appears only to have occurred in full form with the construction of sociocultural cognitive niches in the human lineage. The establishment and maintenance of sociocultural practices—those that make use of public representational systems in particular ways for particular ends—are what accounts for both the initial and continued emergence of content-involving minds.

Only minds that have mastered a certain, specialized kind of sociocultural practice can engage in content-involving cognition. Should creatures with basic minds manage to master such practices, they would gain new cognitive capacities and become open to new possibilities for engaging with the world and other creatures.

Content-involving minds have features and capacities that other, more basic minds lack: they stand apart. This difference can be thought to mark a difference in kind, not just degree, of mindedness in precisely the sense Crane (2014) indicates above. Like Penn, Holyoak, and Povinelli (2008), RECers hold that "Darwin was mistaken: The profound biological continuity between human and nonhuman animals masks an equally profound functional discontinuity between the human and nonhuman mind" (Penn, Holyoak, and Povinelli 2008, 110).

Yet for RECers, the situation is even more complicated than Penn, Holyoak, and Povinelli (2008) make out. This is because REC's distinction between basic, contentless minds and nonbasic, contentful minds does not neatly map onto the nonhuman animal versus human animal distinction. Plenty of human cognition is basic in the sense that it is contentless: this is true not only of human children who have yet to master the relevant sociocultural practices, but it is also true of adults who have—namely, those capable of contentful thought. This follows from the fact that, in the REC picture, even those human minds that become capable of contentful cognition are not *wholly* transformed and thus not fundamentally different from animal minds, *pace* McDowell (1994).[8]

Instead REC holds that the basic character of cognition—including content-involving human cognition—is always and everywhere interactive and dynamic in character. Content-involving cognition need not be content-based; it need not be grounded in cognitive processes that involve the manipulation of contentful tokens.

In an absolutely key respect, on the REC view, human minds—even when content-involving—are like those of all cognitive creatures in terms of their deep, nonrepresentational, interactive nature. This is a non-CIC way of understanding the psychological continuity that exists between contentless and contentful forms of cognition. The difference is that REC conceives of the commonalities present in all forms of cognition in terms of interactive engagement (with the world, with others), as opposed to understanding such commonalities in terms of mental states with representational content. Thus REC assumes that there is a basement-level similarity between all cognitive creatures, and that—*if we focus on those basic features of cognition*—human beings are not cognitively unique.

Nevertheless, despite this important acknowledgment, REC holds that there is a distinction in kind not just between nonhuman animals and humans, but also within the human sphere that is marked by the fact that *only some* minds are capable of content-involving cognition.

REC assumes that content-involving cognition has special properties not found elsewhere in nature and that minds capable of contentful thought differ in kind, in this key respect, from more basic minds. Very well. But does this mean that REC embraces or entails a kind of continuity skepticism? Does REC's insistence that some minds have special, contentful properties that set them apart entail that "we must recognize a sharp discontinuity in the natural history of our species" (Bar-On 2013, 294; see also Fridland 2014)?[9]

In assessing this charge, it is important to disentangle two forms of continuity skepticism that are frequently run together. For example, as formulated by Bar-On (2013), continuity skepticism—of the worrisome diachronic variety we are considering here—boils down to the claim

that there can be *no philosophically cogent or empirically respectable account* of how human minds could emerge in a natural world populated with just nonhuman creatures of the sort we see around us. Few would deny that, biologically speaking, we "came from" the beasts. But the diachronic deep-chasm claim says that we must accept an unbridgeable gap in the natural history leading to the emergence of human minds—or, at the very least, in our ability to tell and make sense of such a history. (p. 294, emphasis added)

Let us bracket, for the moment, the question of philosophical cogency: it needs separate treatment. Rather let us concentrate on the issue of primary concern for naturalists: Does going the REC way in understanding the genesis of contentful minds imply an empirically inexplicable discontinuity in our understanding of the fundamental fabric of the world—one that might incline us to think there is a sort of unbridgeable gap or schism in nature?

Returning to Menary's (2015a) charge, it seems REC is at odds with evolutionary continuity if and only if the following is true: that REC's claim that content-involving cognition differs from basic, contentless cognition in kind and not just degree entails, in key respects, that any explanation of how content-involving minds might have arisen introduces an unbridgeable or at least a scientifically inexplicable gap in natural history from the evolutionary point of view. Does it?

Kinky Cognition: A Sketch of a Possible Story

REC's preferred diachronic explanation of the natural origins of content is kinky. It is kinky because although it doesn't play out along a single dimension—it isn't a simple tale of the mere elaboration of existing forms—it doesn't introduce inexplicable breaks into nature. It does not involve embellishment alone because it does not see all cognitive properties as just more complex versions of what has come before. It is a multi-storey story, one that centrally involves special sociocultural platforms and constructions that enable new forms of cognition. Nevertheless, it is possible to explain how the platforms came to be and how they make those new forms possible without gaps.

REC's tale is one of cognitive niches and how they scaffold more basic minds to introduce novel cognitive features and capacities. The REC assumption is that sociocultural practices introduce something genuinely new and qualitatively distinct into the cognitive mix. Through their acquaintance with culture, some cognitive creatures acquire the capacity to think about the world in wholly new ways. Through mastering what are for them novel practices, they become capable of new forms of thinking of a unique kind.

Does assigning a pivotal role to sociocultural niches in the construction of kinky cognitive capacities put REC at odds with a gapless evolutionary account of the origins of content? That would only be the case if there were some explanatory step between A (contentless minds) and B (content-involving thought) that REC could not account for. If we look closely, from the point of view of understanding our natural history, it is difficult to see exactly where the missing step is supposed to be.

In the abstract, the REC recipe for explaining NOC is fairly simple. A first requirement is that some purely biologically based forms of basic cognition are in place and shared across the species. These are a ground floor requirement and can serve as the first main platform for understanding how and under what conditions contentful forms of cognition could have arrived on the scene.

The positive account on offer draws on, but significantly adapts, some core resources of teleosemantics—the most promising naturalistic theory of content to date—by putting them to a different theoretical use. The teleosemantic apparatus is used to give an account of contentless attitudes exhibiting basic intentional directedness—aka intentional attitudes—as opposed to providing a robust semantic theory of content. This allows us to understand basic cognition in terms of active, informationally sensitive, world-directed engagements, where a creature's current tendencies for active engagement are shaped by its ontogenetic and phylogenetic history.

Basic minds target, but do not contentfully represent, specific objects and states of affairs. Fundamentally, cognition is a matter of sensitively and selectively responding to information, but it does not involve picking up and processing informational content or the formation of representational contents.

Target-focused but contentless intentional attitudes exhibit only Ur-intentionality. Ur-intentionality, as argued in chapter 5, can be explicated by making theoretical adjustments to teleosemantics, thus supplying a fundamental explanatory tool for those working in the enactivist framework. It enables enactivists and others to make a clean and radical break with intellectualist ways of thinking about the basis of cognition and opens up the logical possibility of explaining contentful states of minds as part of entry into sociocultural practices.

To borrow from Dennett (1995), we can take our evolved biological, inherited form of cognition—our first natures—as a starting point for the development of more sophisticated forms of cognition. That is, we can think of evolution as putting in place platforms that act as launchpads, not leashes. Beyond this, for the sociocultural emergence of content, we need to assume that our ancestors were capable of social processes of learning from other members of the species, and that they established cultural practices and institutions that stabilized over time.

It is crucial to REC's NOC story that biological capacities gifted by evolution could have given rise to social learning. So far so good: surely, there is nothing mysterious or gappy on the table yet. The capacities in question can be understood in biological terms as mechanisms through which basic minds are *set up to be set up* by other minds and *to be set off* by certain things.

As a baseline, all that is required is that basic minds can target and respond to certain things in ways relevant for learning various tasks from their fellows. According to REC's teleosemiotic view, individuals can respond to the environment and to each other in ways that allow for emulation, imitation, and regulation of what they target and attend to in ways that make basic forms of social learning possible.[10]

There is no reason to suppose that the cognition at play in such social engagements and interactions must be grounded in representationally based rules of any kind. Rather, all that needs to be assumed is that normally developing participants in such practices are already set up, nonaccidentally, to target and tune into the expressively rich intentional attitudes of others.

None of the basic cognitive activity described above requires any purposive rule compliance on the part of participants. Positing mechanisms of social conformity would suffice to explain how creatures with only basic minds could come to be set up by others and to set up others. If such an account is tenable, it could explain how the practice of social learning might get off the ground in an evolutionarily respectable way without bringing any contentful attitudes into the story.

Critics have argued the REC story, as sketched above, won't fly. But this is not because it is evolutionarily unsound; indeed, quite the contrary, it is because REC's naturalistically respectable resources are too crude to tell the story properly. The REC story is gappy, not because it introduces kinks or evolutionary gaps, but because it can't fill in all the relevant details. This, so its critics claim, is because REC's vision of contentless basic cognition is just too meager and weak to serve as a tenable foundation to account for the relevant action.

This negative assessment is based on the repeated accusation that REC's content-free account of basic minds reduces to, or has no more resources than, a crude stimulus-response form of behaviorism (O'Brien and Opie 2015; Kiverstein and Rietveld 2015).[11]

Should the "mere stimulus-response behaviorism" characterization turn out to be true of REC, then its Radically Enactive, Embodied account of basic Cognition would really be nothing more than a Radically Enactive account of Behavior—REC would be REB, after all. If REC were REB it would clearly lack the resources for supplying a biologically credible story about how creatures, with only contentless states of mind, could have engaged in flexible kinds of basic and social cognition of the sort needed for triangulating in primitive, nonlinguistic ways with each other.

If the "REC is really REB" analysis proves correct, there is no prospect for REC to give the sort of explanation, as sketched above, of how children could have become players in the relevant sociocultural games without assuming the existence of some kind of prior contentful mentality. REC would have no real chance of accounting for even the most basic forms of social cognition.

Making this latter worry explicit, Lavelle (2012, 469) argues that REC, explicated only in terms of intentional attitudes directed at natural signs (à la Hutto 2008), encounters special problems when it comes to explaining "the flexibility of prelinguistic social interactions." If Lavelle's diagnosis is right, it suggests that engaging with others to form even the most primitive intersubjective triangles –of the sort exemplified by nonlinguistic joint attention– requires all participants to be in and to ascribe contentful states of mind. Naturally, these observations, if true, would lend support to the claim that "mindreading is required to learn language in the first place" (Carruthers 2011, 321).[12]

Should the "REC amounts to REB" charge hold up, then the possibility of providing a credible REC account of the emergence of content along NOC lines, let alone a gapless one, can be pretty safely ruled out. The big question is: Does the charge hold up? Consider Lavelle's (2012) claim that REC only gives an account of basic cognition in terms of natural signs. Although an account of how basic minds respond to natural signs in hardwired ways is part of the REC story, it is not the whole REC story.

RECers nowhere claim that contentless ways of responding to worldly offerings—including the attitudes of others—must be simple, fixed, or slavishly automatic. This is clearly not true even of our most basic, most infantile ways of responding to and engaging with others. In most cases in which Mother Nature initially fixes intentional targets, it is perfectly possible that how we respond to such targets can be shaped in dynamic, spontaneous, and context-sensitive ways (see chapter 5, and also Hutto 2006a, 2006b).

Exactly how such enactive engagements will unfold in any particular situation will depend on a number of hard-to-predict factors about what is targeted and how it is targeted. The pattern of current responses depends on individuals' past interactions with similar situations and the unique features of the current context. Such factors make a difference not only to the intentional attitudes of individual participants but also in how they respond to the manifest intentional attitudes of each other. Yet there is no need for the participants in such engagements (or some subpersonal cognitive system within them) to be aware of, or have knowledge of, or represent the steps of these dynamic processes in order to respond flexibly to situations.

That REC does not reduce to REB can be clearly seen in the way it allows that basic minds can have quite flexible and expressive modes of operation. For example, REC is perfectly compatible with, and embraces, Bar-On's (2013) account of expressive attitudes given in terms of animal signs that are not mere natural signs.[13]

By Bar-On's (2013) lights, expressive attitudes of the kind that feature in prelinguistic triangulation are not to be understood in terms of natural signs precisely because responses to natural signs alone would not account for the fact that animal mental capacities are deployed in interesting and flexible ways.

Repsonses to animal signs cannot be understood as mere triggered reactions to the reliable indication of proximal stimuli. Rather they are subtle and adjustable responses to sophisticated patterns of expressive behavior—understanding them requires getting a grip on "complex networks of animal communication" (Bar-On 2013, 300). The richly expressive behaviors in question include "yelps, growls, teeth-barings, tail-waggings, fear barks, and grimaces, lip smacks, ground slaps, food-begging gestures, 'play faces' and play bows, copulation grimaces and screams, pant hoots, alarm, distress, and food calls, grooming grunts, … and so on" (Bar-On 2013, 317).

Sensitive responses to such expressive behaviors can be fully understood in terms of contentless intentional attitudes being directed at the contentless but expressive intentional attitudes of others. As Bar-On (2013, 320) makes clear,

Acts of expressive communication often involve an overt gaze direction, head tilt, or distinctive bodily orientation guiding the receiver's attention not only to the expressive agent's affective state but also to the object of that state—the source or target of the relevant state. [For example,] ... a dog's cowering demeanor upon encountering another will show to a suitably endowed recipient the dog's fear (kind of state), how afraid it is (quality/degree of state), of whom it is afraid (the state's intentional object), and how it is disposed to act—for example, slink away from the threat (the state's dispositional "profile").

The important thing to note about expressive signs is, as Medina (2013, 326) highlights, that they should not be understood

on the Gricean model of conventional signs, that is, as involving or requiring fully formed communicative intentions and internal representations. Expressive behavior is not self-reflective intentional-inferential communication among rational agents who are representing each other's minds and their contents. The production and the uptake of expressive behavior place much weaker representational demands on their producers and responders than self-reflective intentional-inferential communication does. ... [Even so,] expressive acts have a significant degree of spontaneity that distinguishes them from automatic physiological reactions.

These observations are revealing in two crucial respects. First, they show how in offering its particular account of contentless minds and Ur-intentionality REC does not reduce to REB. It should be clear now that the "nothing but REB" charge is a complete red herring: it seriously underestimates and mischaracterizes REC's resources. Second, it should also be clear that it is entirely possible to offer alternatives to mindreading explanations of basic social cognition and primitive triangulation that do not bring mental contents into the picture at all, *pace* familiar CIC advertisements.[14]

In sum, once REC is properly characterized, there is no reason to doubt that it is well positioned to explain the origins and basis of noncontentful forms of intersubjective triangulation (see

Hutto 2011, 2015). With these pieces of the puzzle in place, it should also be clear that there is nothing in REC's assumptions about basic minds that ought to make its NOC story about the emergence of content evolutionarily implausible. The forms of social interaction it requires are possible without presupposing any contentful attitudes.

That takes care of the foundation. But how, when, and where does content enter the picture? Content only arises when special sorts of sociocultural norms are in place. The norms in question depend on the development, maintenance, and stabilization of practices involving the use of public symbol systems through which the biologically inherited cognitive capacities can be scaffolded in particular ways.

The practices in question are claim-making practices—and they are special because they require participants not only to respond to things but to do so by *representing them as being thus and so* independently of what might be said about them. The claims in question must satisfy what McDowell (1998, 222) identifies as the "familiar intuitive notion of objectivity" that is bound up with a "conception of how things could correctly be said to be anyway—whatever, if anything, we in fact go on to say about the matter."

Only with the appearance of such claim making does it become possible to make the special kinds of semantic errors unique to contentful thought and speech. Getting things wrong in a truly representational sense is not just a matter of being literally misguided in the way purely biological entities and creatures can be. It involves being subject to the censure of others—not just in the sense of being in or out of line with what is acceptable or not for some community, but being able to get things right or wrong in a game in which it is at least possible to be right according to how things are anyway.

Only those in a position to play this sort of game can be said to have content-involving thoughts and speech. And those capable of playing such games have an enhanced cognitive repertoire that is of a quite different order and kind than is deployed in only basic cognitive operations. Despite these special features, the emergence of content-involving cognition can be explained in naturalistic terms, without residue, by appeal to the way— with the right support from others— biologically basic cognitive capacities get used under the right conditions.

REC assumes that the normative practices required for claim making arose with the advent of special kinds of practices made possible by the establishment of sociocultural niches.[15] In assuming this, REC's NOC story follows Clark 2006a in supposing that content-generating practices are part of a "cognition-enhancing animal-built structure ... a kind of self-constructed cognitive niche" (p. 370; see also Clark 2006b). From a naturalistic point of view, unless this assumption about cognitive niches is at odds with evolutionary continuity, it is difficult to see what in this sketch of a REC-inspired NOC story entails a discontinuity in nature.

The trick to understanding the emergence of content is to understand the emergence of a special sort of normative sociocultural practice involving the use of public symbols. Thus unless there is something deeply mysterious about social conformity and cultural evolution, there is nothing in the proffered explanation that introduces any inexplicable gap into nature.

7 Perceiving

Does it follow from the sense impressions which I get that there is a chair over there?—How can a proposition follow from sense-impressions? Well, does it follow from the propositions which describe the sense-impressions? No.—But don't I infer that a chair is there from impressions, from sense-data?—I make no inference!
—Ludwig Wittgenstein, *Philosophical Investigations*

Out of the Armchair

Some thinkers fervently believe that perceiving always entails the existence of perceptual contents—that perceiving entails the existence of some kind of correctness conditions, such as truth, accuracy, or veridicality conditions. The assumption that perceiving must involve representational content in this sense is usually bound up with the conviction that any instance of perceiving must involve taking or depicting the world to be a certain way such that it might not be that way.

How might one defend such an essentialist thesis? The most straightforward way is to claim that it is an analytic truth, known by some a priori means.

In rejecting REC's suggestion that it is possible to think otherwise about perceiving, Campbell (2014, 175) insists that the very idea that perception might be contentless is "manifestly implausible." He attempts to lend force to this conclusion by reminding us of the allegedly essential properties of perceiving. Drawing on his own experience, he assures us that: "When I gaze at the Müller-Lyer illusion, I see one line as being longer than another, when actually the two lines share the same length. My experiences thereby present things as being a way they are not: that is, they (surely) misrepresent reality. *So* ... perceptual experience is representational" (p. 175).

Ultimately, this analytic style of argument by attempted demonstration appeals solely to intuitions about cases. It assumes that we can settle substantive debates in philosophy by reviewing what we find obvious about a case or by consulting our intuitions. In this instance, Campbell's exercise is supposed to suffice to show that it is logically impossible that one can see a line as being longer than another without representing said line to be longer. The intuition pumped, as Dennett (1995, 2013) would say, is meant to rule out the possibility that things can look a certain way to perceivers without it being the case that they represent things as being a certain way. Yet despite the power this intuition holds for some, there is no obvious contradiction in assuming that things can look a certain way to perceivers—as revealed in their responsiveness to objects—without such perceivers contentfully representing things to be any way at all (for a discussion see Hutto and Myin 2013, 121ff.).

In reply to this suggestion Campbell continues to reject the possibility that REC seeks to promote, insisting that there is an entailment relation here that leaves REC no logical room for maneuver—a fact RECers apparently fail to notice. REC's position on perceiving is thus deemed "to put it mildly—a difficult position to sustain. If my experiences are such that 'things look and feel a certain way' to me, but things are not the way they look and feel, then in what senses of the words 'inaccurate' and 'non-veridical' are my experiences otherwise than inaccurate and non-veridical?" (Campbell 2014, 175).

The answer, of course, is that there is an important difference between something's looking or feeling a certain way and its being taken to be a certain way. Nevertheless, we don't expect to persuade anyone who doesn't see this by tracing over the contours of such examples—certainly not when deeply entrenched intuitions frame the way the participants in this debate think about the relevant cases.

In any case, card-carrying naturalists, as argued in chapter 1, should be wary of trying to rule out the very idea of contentless perceiving from the armchair. The evolution of debates around this very topic provides a salutary lesson of the dangers of making unshakeable a priori pronouncements about what essential properties perceiving must have based solely on what our intuitions tell us has to be the case.

As Brogaard (2014, 1) observes in the introduction to her edited collection *Does Perception Have Content?*, only a few years ago the titular question posed by that work "would not even have been considered. Perhaps it would not have seemed intelligible. But things have changed, and there is now a considerable number of articles, theses, and books aimed at answering it, positively or negatively."

What is responsible for this sea change? Brogaard (2014) surmises that it has been driven in large part by new developments in naturalistic philosophy of mind, developments centered around the very issues about how best to think about cognition for scientific purposes that we have been exploring in this book—issues tied to debates about whether the future of cognitive science lies with representational or nonrepresentational approaches to cognition. The only properly naturalistic way of conducting such debates is to stay firmly focused on what the explanation of phenomena requires.

Of course, the background condition for having any meaningful debate is agreement about the terms in dispute. We can't assess whether we should think of perceiving as involving content or not if we can't settle on what it means for perceiving to involve content—that is, if we can't settle on what is being proposed or denied by various parties to the discussions.

As we have done throughout, we understand representational content in a broad way, taking it to imply that representing the world contentfully is a matter of taking ("representing," "saying," "asserting," etc.) things to be a certain way such that they may or may not be that way. We take it to be uncontroversial that this is the default, indeed the textbook way, of thinking about the nature of representations.[1]

Bearing this in mind, this chapter aims to assess whether positing representational contents understood in this canonical sense gives a real explanatory punch to theories of perceiving, or whether positing them is only a distracting gloss that generates imponderable mysteries for both science and philosophy.

Once More unto the Predictive Breach

The hottest and most promising new developments in perceptual science are surely those theories that make heavy use of Bayesian decision theory in modeling perceptual processes. Predictive processing accounts of perceiving require us to do a

180 in shifting our perspective away from the old school, classical cognitivist vision of perceiving. That classical view—canonically formulated by Marr (1982)—focused on trying to explain how in perceiving we build up more and more complex representational models of the world based only on information transduced from various sensory receptors.

We discussed the Marr-style traditional take on perceiving and related philosophical accounts supporting it—those supplied by Fodor (1983) and Burge (2010)—and revealed the shortcomings of such proposals in our previous work (Hutto and Myin 2013, chaps. 5, 6). In this chapter we go a different way, taking a second look at PPC accounts of perceiving—the proposed successor to the classical cognitivist framework—building on our initial attempt in chapter 3 to sketch the fundamentals of PPC and a REC rendering of it. We think that there is something right about the PPC framework and that it offers a better approach to perceiving than the neurocomputational theories of classical cognitivism, but only if it is interpreted properly. Since we think the proper interpretation of PPC requires a move away from representationalism, it is worth revisiting the debates around that topic once again to further motivate going the REC way.

Many philosophers maintain that a representationalist interpretation of PPC is simply unavoidable and nonnegotiable. The fundamental use that PPC makes of Bayesian rules and theories, they say, assigns "a central explanatory role to representational content" (Rescorla 2016, 24).

Accordingly, if this is an unalterable fact about the PPC framework, and if that framework proves to be the most promising basis for understanding perceiving, then we have a positive reason to "embrace representation as a genuine, scientifically indispensable aspect of mentality" (Rescorla 2016, 24). Should all of these claims prove true, the representational pretensions of many standard renderings of PPC will turn out to be entirely justified a posteriori, just as Gładziejewski (2016) claims.

To be utterly clear, and to avoid any accusation of setting up a strawman, the PPC interpretation we are considering operates with the very understanding of representational content that RECers target —the generic notion used at large in the classical cognitive science literature. The notion of representational content in dispute assumes the existence of correctness conditions of some kind, which can be variously construed as truth, accuracy, or veridicality conditions.

For example, Rescorla (2016) makes it indelibly clear that the notion of representation he takes to be in play in PPC accounts has this key property. He writes:

Philosophers and scientists use the phrase "mental representation" in many different ways. The type of mental representation that concerns me involves representational content ... [such that] a mental state has a content that represents the world as being a certain way. We can ask whether the world is indeed that way. These states are semantically evaluable with respect to such properties as truth, accuracy, and fulfillment. (p. 17)

To evaluate whether representations so understood are in fact indispensable to PPC, we need to be clear about why such philosophers think representations are necessary to PPC's account of the working parts of cognition. Most of those involved in this debate do not give any airtime to the hyperintellectualist view that representational contents are needed to represent the principles involved in perceptual processes.[2] Of course, everyone who accepts PPC agrees that the principles of perceptual processing conform to, and can be described in terms of, Bayesian laws, but there is no need—and indeed no advantage—

Perceiving

in characterizing the brain as literally making Bayesian inferences concerning such principles when carrying out its anticipatory work.[3]

Instead it is argued that a representational reading is needed to understand the many multilayered and multiscale intermediate steps and stages of perceptual processing. It is here that top-down hypotheses and bottom-up data meet and the brain employs weighted precision estimates at this interface as part of its effort to make its best guesses about what it perceives in the world. Thus it is in the ebb and flow of perceptual processing itself that, so it is claimed, we find "a rich structure of representational mental states that mediate between sensory stimulation and bodily motion" (Rescorla 2016, 24). Since perceptual processing operates on many levels, layers, and scales all at once and since the products of each level are used at other levels in the hierarchy, representationalists claim that "representational mental states figure crucially as both explanantia and explananda within our current best sensorimotor psychology" (Rescorla 2016, 23). Despite the fact that, as Rescorla (2016, p. 20) reports, PPC remains open about the precise nature and exact content of the brain's hypotheses, "it makes one crucial assumption: each hypothesis is accurate or inaccurate, depending on the environment. Thus, each hypothesis has an accuracy-condition." (p. 20)

Gładziejewski (2016, 565) tells a similar story. He emphasizes the need to posit contentful representations in order to allow for the potential mismatches that occur during prediction error minimization—mismatches between how things are and how the brain/mind "represents" them as being.

The PPC assumption is that the more accurate the brain's generative model is—in terms of its likelihoods, dynamics, and priors—the more accurate the brain's hypotheses will be about the causal-probabilistic structure of the external world. It is this reading in terms of the content of the brain's bets made during perceiving that makes it seem unavoidable that representational contents of a quite traditional kind must play a crucial role in PPC explanations.

In illustrating the point, Gładziejewski (2016, 575) imagines a case in which a human brain uses a less-than-accurate generative model of the world and thus "settles on the hypothesis that it is seeing a plush imitation of a tiger ... when what it in fact faces is a live tiger." Getting it wrong in this sort of case has fairly obvious costs. And even if the brain has no direct perceptual access to tigers, we can expect that when "what one is in fact observing is a tiger, then the hypothesis that the inflow of sensory information has been caused by a tiger will generate (on average) a smaller prediction error than alternative hypotheses—including any that attribute the causal origins of the incoming signal to a plush toy or a domestic cat" (Gładziejewski 2016, 562).

It should now be clear why anyone who understands perceiving along PPC lines is likely to think that PPC is fundamentally committed to positing processes that involve the manipulation of representational contents.

No doubt adopting a realistic representational reading of PPC is the easiest and most natural way to understand that framework. But taking stock of several gaping holes in the theoretical foundations of representationalist renderings of PPC might encourage its fans to consider the possibility of an alternative, nonrepresentational formulation of the framework along the lines REC offers.

What holes? As we saw in the discussion of bootstrap hell in chapter 3, if we accept a cognitivist reading of PPC, then it appears that we have no good answer to the question: Where does prior knowledge—the content of initial hypotheses—come from?

Nor do we know "how the nervous system implements Bayesian activity" (Rescorla 2016, 13). As discussed in chapter 2, when this lacuna is understood in light of the HPC we may become suspicious of familiar attempts by cognitivists to play their standard card of claiming that such a gap in our knowledge doesn't matter because "one may legitimately describe mental activity at an abstract level that prescinds from neural implementation details" (Rescorla 2016, 13).

Most fundamentally of all, as discussed in chapters 2 and 5, and as Rescorla (2016, 17) admits, we don't yet have an account of "how mental states come to have representational content".

In the end, if this were all representationalists had to offer on the theoretical front, those working in the field might well be persuaded to explore a REC reading of PPC—as initially sketched in chapter 3—in their quest for a naturalistic account of what does the real work in basic perceiving.

Unfortunately for everyone, a REC reading of PPC is allegedly already ruled out a posteriori, if Rescorla (2016) is right. Apparently, as matter of fact, the science says no. Indeed, Rescorla (2016, 18) is quite fond of reporting what the science says or does, and of deploying "the science" to "rebut various views."[4]

Even so, he is also quick to acknowledge that scientists don't actually make any pronouncements of the kind he reports about the science's central theoretical commitments. Rather, he tells us:

Admittedly, sensorimotor psychologists do not use locutions such as "representational content" or "veridicality-condition." Nevertheless, such locutions illuminate the explanatory structure of optimal control sensorimotor modeling. Here again there is room for fruitful exchange between philosophers and scientists. Tools from philosophy of mind can help sensorimotor psychologists better understand the conceptual foundations of their own research. (Rescorla 2016, 23)

So the claim that PPC cannot possibly tolerate a REC reading really boils down to the claim that if the science understood its philosophical foundations correctly—namely, in representationalist terms—then it would say no to REC. Despite Rescorla's firm, ex cathedra insistence about what "the science" does or should say, all this looks a bit circular from where we are standing.[5]

Of course, should defenders of representationalist readings of PPC be able to answer questions about its neural implementation, show how it is possible to avoid bootstrap hell, or deal with the HPC, then the path of least resistance for all involved would be to simply go cognitivist about PPC.

Gładziejewski (2015) tries to do better than most on this score, making an effort to show how it is possible to pay for the representational content that is putatively indispensable for PPC explanations. His proposal is of special interest precisely because it does not rely on teleosemantics, which, if the analysis of chapter 5 holds up, is off the table.

Gładziejewski (2016) advances the claim that PPC should postulate internal representations whose functional profile is nontrivially similar to the functional profile of cartographic maps. He identifies four features that qualify such maps as representations, the first of which is crucial to the positive story of content he hopes to tell. Allegedly, maps represent by

1. Structurally resembling features of some domain
2. Guiding the actions of their users
3. Doing so in detachable ways (e.g., they can be used "offline")
4. Allowing their users to detect representational errors[6]

Drawing on work by O'Brien and Opie (2004), Gładziejewski (2016, 566) proposes that maps "represent in virtue of sharing, to at least some degree, a relational structure with whatever they represent."[7]

Explicating this idea, O'Brien and Opie (2015) conceive mental representations in structural or analog terms—namely, in terms of physical analogies that hold between a representational content and what it represents. On this view, representational contents are just intrinsic structural properties of representational vehicles. More precisely, mental representations are contentful in virtue of the fact that they share resemblance properties of some structural kind with what they represent. The content of an analog representational vehicle is thus fixed solely by the structural resemblances that hold between the vehicle and its object. The resemblances in question are thought to depend only on intrinsic properties of the vehicle and objects to which the vehicle relates.

A structural resemblance theory of representational content will be attractive to anyone who is convinced that other naturalized theories of content face intractable problems. However, there are good reasons to think, as discussed briefly in chapter 2, that without further theoretical backing, simply positing stand-ins that only structurally resemble items in a given domain can be better understood in entirely nonrepresentational, noncontentful terms.

To see this it helps to get clear about the general character of stand-ins. Consider a mundane case. Someone, Y, acts as a surrogate for someone else, X, by delegating for X and playing X's role at some function or event. Clearly, even though we might say that Y is X's representative, it is clear such surrogacy is achieved without X actually representing Y in any contentful sense. We can also assume that in order to play the surrogacy role on some occasion, it is further required that Y structurally resembles X. Perhaps Y needs to structurally resemble X in several respects at once. Allow that some of these respects are quite abstract, such as sharing X's height or X's capacity for witty repartee. Even in such cases Y does not represent X in any contentful sense.

As it is with one person acting as a surrogate or stand-in for another, so it is with a cognitive state or process that acts as a surrogate or stand-in for another. This is so even when the states or processes are required to structurally resemble their counterpart states in quite abstract ways. The moral is that being a stand-in that structurally resembles another state or process does not, by itself, explain how or why such stand-ins should be thought to instantiate or bear representational content. Of course, if the stand-ins in question are already content-bearing, then that's a different story. But then we need to be told where that content comes from, presuming it is not simply presupposed that the stand-in in question bears content.

Stand-ins that structurally resemble external items in various ways might be used to play specific causal roles in mapping some domain so as to systematically guide performance and behavior. Still, even if all of these conditions are met, there is no reason to suppose that such stand-ins qualify as content-involving representations.

What this analysis reveals is that structural resemblance theories of content don't suffice, even when structural resemblances are exploited for tasks such as mapping a domain and guiding behavior with respect to it. Additional, independent reasons are needed to explain any content such stand-ins might bear.

Having come this far, a different puzzle arises for friends of structural resemblance theories of content. For, even if surrogates do happen to bear contents, exactly what causal or explanatory work could such contents do in the story of perceiving? What role might they play in cognition that the other properties—the structures that bear them—do not already usurp?[8]

Given all of this, it is hard to see why even cheaper contentless alternative accounts of stand-ins and surrogates—those that only assume structural resemblance properties—are not best placed to do all of the relevant cognitive labor without generating any unwanted, intractable mysteries.

Certain correspondences must hold between a map and what is mapped in order for a map to be used nonaccidentally to successfully navigate some environment or domain. But per the analysis provided in chapter 5, functional isomorphisms are all that need to be exploited for the purposes of mapping and navigating. Yet it is not at all obvious why the exploitation of such correspondences need entail the existence of representational contents.

In sum, it seems that all the explanatory work can be outsourced to less costly employees than contentful representations when it comes to explaining the sort of mapping phenomena that Gładziejewski (2016) thinks are central to a PPC account. REC is entirely open to the possibility that the full story of such cognition might be told in terms of items that bear structural resemblances to other items and which are actively exploited in systematic ways in order to enable organisms to bring their actions to bear on teleologically fixed targets. If that is all that is needed to do all the work in explaining a system's adaptiveness in PPC terms, then there is no need and no gain, only pain, in insisting that we must posit representational contents to make sense of the mediating processes involved in cases of basic perceiving.

Yet, for all that has been said, some theorists hold that there is another place where representational contents might need to show up in a PPC story. For even if it is agreed that representational contents are not needed to represent the principles of perception and that they are not needed to explain perceptual processing itself, it might still be argued they are needed for understanding the products of perceptual processing.

This is precisely the line of thought Orlandi (2014) defends. She agrees with REC in holding that there is no need and no advantage in characterizing the brain as literally making contentful inferences when carrying out its perceptual processing work.[9] Thus she holds that

we do not need to appeal to representations to explain what vision does. ... We can explain the central phenomena that we need to explain by thinking of the visual process as mediated by functional states and features that are *better understood non-representationally*, and making reference to environmental conditions, in particular to statistical regularities in the world with an eye to organismic needs. (p. 3, emphasis added)

Despite arguing that the processing of the visual system can be fully explained in terms of the way it adapts to environmental regularities, Orlandi (2014) stops short of completely going the REC way. Instead she claims that products of perceptual processing are always representational, that perceptual processes issue in states that are representations "because they have all the … basic attributes of representational states" (p. 13). Moreover, following established lines of thinking, she identities the content of perceptual representations with their accuracy conditions (see Orlandi 2014, 10).

The products of perceptual processing must be contentful in the familiar representational sense, Orlandi (2014) holds, because their jobs are to stand in for (p. 9) or stand for (p. 19) things that are not "strictly speaking, present to the senses" (p. 31). Here she invokes the familiar line of argument, advanced by Clark and Toribio (1994), about the need to posit representations in cognition whenever faced with representation-hungry tasks involving what those authors iconically dub "the absent" and "the abstract."

We have already provided reasons in chapter 2 for being skeptical about this general style of argument. Still, it is worth taking a second look, focusing on some perceptual examples that impress Orlandi (2014) and motivate her to endorse her particular brand of representationalism.

Orlandi (2014) wonders how the visual system can be systematically influenced by the backsides of objects, which are strictly unseen, when it perceives objects as wholes. Clearly, although the backsides of things are present in such cases, they are also hidden from sight in that they do not directly impinge on our senses. Despite making a difference in acts of perceiving, the hidden parts do not directly stimulate or provide feedback to the eyes (see Orlandi 2014, 125).

By the same token, Orlandi wonders how constant and stable properties of an abstract sort manage to influence the visual system despite being technically unperceivable. For example, despite the fact that the visual systems might receive only electromagnetic radiation of, say, a cow, it still is able to perceive the abstract property of "cowness" (see Orlandi 2014, 127).

In these kinds of cases Orlandi (2014) holds that some token perceptual states that are products of perceptual processing must stand in for the sorts of items that cannot be immediately and directly perceived in the environment. It is such stand-ins, she holds, that fill out our perceiving and guide further cognitive processes and action. Such representational states are thought to contentfully inform other states and guide performance precisely because the absent and abstract items they stand in for cannot play such roles. Orlandi's (2014) general verdict about what is required to tell the whole story of perceiving in such situations takes what is, by now, a familiar form. Thus she holds that the "later visual states, the 'outputs' of visual processing, are plausibly representations, because they inform actions that are stimulus-independent" (p. 62).

Yet, as argued above, there is no compelling reason to think that contentful representations are required even if stand-ins substitute for strictly unperceivable items and thus exert a systematic influence on behavior in cases of basic perceiving.

Strikingly, sometimes Orlandi (2014) is sensitive to this fact and sometimes not. Her official story about the need to posit representations in perceiving is inconsistent on this issue. It fails to treat like cases alike. Thus if her reasons for construing the products of perceiving as representational are good reasons, then they apply just as forcefully to what she says about the processes of perceiving. And if those reasons are not good reasons, then the opposite holds.

Consider that Orlandi argues, as does REC, that positing sensitivities to statistical patterns of variation is all that is needed in order to explain perceptual processing and that there is no need or advantage in positing contentful representations in such explanations. Yet such statistical patterns are just as unavailable to directly inform perceptual processing as are the backsides of objects are to inform the final products of visual experiences. Moreover, the statistical patterns in question are entirely abstract. Thus if Orlandi's reasons for thinking that visual processing is not best explained in representational terms are sound, then they are also reasons for thinking that the products of perceiving are not best understood in representational terms. As a matter of theoretical elegance and consistency, these verdicts should stand or fall together.

It is easy to see how Orlandi's argument for nonrepresentationalism about perceptual processes can be converted into an argument for nonrepresentationalism about perceptual products. This only requires making two simple substitutions—replacing the terms *inferential* and *processes* with *content* and *products* respectively—in the relevant passage from Orlandi 2014:

> It seems that despite their plausibility, [content] views are not explanatory. They seem blind to what is external to the system—for example the fact that the stimulus itself has typical causes. Because of this apparent blindness, they also have a tendency to over intellectualize perceptual [products], and to think that visual systems need internal representational resources. (p. 41)

Of course—for all we've said above—one might go the other way on this issue to secure theoretical elegance, endorsing representationalism in both cases.[10] Yet if the considerations of the early sections of this chapter as well as previous chapters are sound, they provide reasons for thinking that the scales tip in favor of adopting a thoroughgoing nonrepresentationalism about perceiving when it comes to understanding both its processes and products.

Integration and Interface

It should be clear that if representationalism is not the best option for thinking about perception within single modalities, such as vision, then there is probably no good reason to think that representations are needed to explain how different perceptual modalities interact with one another or even multiple others at once. Nanay (2014, 46) argues otherwise:

> Multimodal perception seems to require matching two representations, a visual and the auditory one. If we cannot talk about perceptual representation, how can we talk about what is being matched? The auditory sense modality gives us a soundscape and vision gives us a visual scene and our perceptual system puts the two together. It is difficult to explain this without any appeal to representations. The enactivist arsenal seems insufficient: they can appeal to the active exploration of the multimodal environment, but this is unlikely to help here: we are actively exploring the world that is given to us in both sense modalities—but this in itself requires multimodal integration. In short, the active exploration of the environment presupposes multimodal integration, which, in turn, seems to presuppose representations.

In an effort to strengthen his argument, Nanay (2014) provides an example of such multimodal integration, citing a study by Shams, Kamitani, and Shimojo (2000) in which perceivers presented with only one visual flash are caused to visually experience two flashes when they also hear two beeps when the visual flash is shown.

According to Nanay (2014), the intermodal interaction at play in such demonstrations cannot be explained in REC's way. This is because, he reasons, any explanation given in terms of the active exploration of the environment—and how such exploration shapes perceivers' expectations—must already presuppose the existence of intramodal integration. Fair enough. Let us agree that this is so. Still, it is hard to see how—without blatantly begging the question against REC—it follows from these facts alone that we can only explain multimodal integration by appeal to contentful representations. All we are told is that the matching required in this kind of integration demands, or "seems to" demand, a representational explanation.

REC explains the patterns of responsiveness found in intermodal integration in just the same way it explains the patterns of responsiveness found in intramodal perception, namely by invoking the idea that perceivers adapt to naturally occurring regularities acquired during their history of previous interactions with the environment. In the intermodal case these regularities concern co-occurrences between stimulation in more than one modality. Given regular co-occurrences between sights and sounds, and adaptation or attunement resulting from dealing with such occurrences, organisms become adapted to responding to intermodal regularities in characteristic ways. Through such encounters organisms are set up to respond in predictable ways to multimodal stimulations—a process that includes acquiring specific tendencies for brain connectivity between various modalities.

This history of interactions explains why an organism would tend to perceptually experience two flashes when it hears two beeps, while strictly speaking, it sees only one flash. As Shams (2012) observes, it is adaptive for organisms to rely on the combined stimulation of different modalities rather than single modalities in isolation.

Appeal to such facts suffices to explain why the intermodal unification—the fusing of the sightscape and the soundscape—happens in the first place, and why it happens the way it does. It is unclear from what Nanay (2014) says why a special kind of matching occurs in this process or why contentful representations would be needed to explain it. That is to say, it is unclear why more than appeal to the way organisms are shaped by their interactional histories is needed in order to explain the capacity for multimodal integration.

Still, some may have the nagging thought that the above account still misses something important – that something more is required to explain the synchronic integration of various cognitive phenomena, and that something must come in the shape of mental contents. A subtly persuasive thought in the field is that mental contents are needed not to explain how cognitive phenomena manage to integrate and interrelate per se, but how they do so intelligibly.

The central posits of cognitive theory—informational contents and contentful representations—are specially tailored to explain intelligible integration and interface. Mental contents are meant to provide the perfect theoretical glue for understanding the allegedly meaningful character of cognitive integration. Cognitive neuroscience promises to show how there can be relevant connections between various mental phenomena in a way that provides explanations of cognitive integration that cannot be understood in terms of mere correlations or brute causal relations.

It is the apparent need to make room for this special kind of intelligible integration that forms the basis for yet another argument that cognitive science cannot get by without calling on mental representations in its explanations. Hence, the standard assumption in the field is that information processing mechanisms that underwrite cognition are at once both causal-mechanical and intentional in character. A marriage of these properties is, however, only possible if it is assumed, contra REC, that minds do, in fact, traffic in informational contents and representational contents.

Representationalism must be true, so this line of thinking goes, because it is only by assuming its truth that cognitive science can explain how it is possible to blend the intentional with the neural, thus accounting for cognition's unique properties—the very properties needed to explain how cognitive phenomena can intelligibly integrate.

Thus the solution to the intelligible integration problem requires embracing a cognitivist vision of mind—one "organised as a hierarchical system ... which *uses representations* of the world and its own states to control behaviour" (Gerrans 2014, 47, emphasis added).

Representationalism is the panacea because it allegedly confers unique explanatory advantages that will put us in a position to "show how *facts* identified and explained by disciplines operating at 'levels' such as molecular neurobiology or neuroanatomy *can explain* psychological and phenomenological level *facts*" (Gerrans 2014, 20, emphasis added).[11] By these lights, what makes a cognitive theory pitched at the information processing level so uniquely valuable is that it allows us to "*bridge the gap* between neurobiological and personal level explanation" (Gerrans 2014, 21, emphasis added).[12]

In seeking to show why such a theory is needed, Gerrans (2014) contrasts his integrationist proposal with the view that the mental is an autonomous domain forever set apart from what happens in the brute, neural domain, to which it is linked only by mere causal relations.

The autonomy thesis is most often defended on the grounds that "rationality is essential to mindedness" (Graham 2009, 12). The most familiar argument for the idea that what is properly mental is irreducibly autonomous is that rationality is the hallmark of the mental—it is a feature that sets the properly mental forever apart from all other kinds of phenomena.[13] Mental phenomena with the relevant forms of rationality, on this view, exist only in the space of reasons.

The autonomy thesis has been accused of promoting a disunified framework when it comes to thinking about minds—one in which mechanisms are assumed to make only a causal difference to cognitive goings-on in a way that debars them from being properly explanatory (see Gerrans, 2014, 15, 20).[14]

To avoid this supposed disunity problem, understanding the mind through the lens of cognitive theory is meant to inject a peculiar sort of intelligibility into the mix—a kind of intelligibility that allows theorists to go beyond the telling of mere difference-making causal stories. Cognitive theory allows us to see how all things cognitive fit together in a systematic way; it bridges the gaps and enables explanations at many different scales and levels to be intelligibly connected by detailing how information flows from level to level and what role particular processes play in the wider cognitive economy (Gerrans 2014, 48; see also pp. 32, 53, 79, 103). Only the contentful and representational entities posited by cognitive theory, so it is claimed, have what it takes to do the required integrating work.

Revealingly, if taken seriously cognitive theory provides us with a new mark of the mental—not one that assumes rationality is key, but one that assumes content is key.[15] For even though, on the surface, it looks like a mixed bag of things fall under the category of the cognitive—and even though cognitive phenomena span the subpersonal and the personal—they are able to integrate intelligibly because they all involve the processing of information and the manipulation of representational contents. This is what truly defines minds.

We don't deny the attraction of positing mental contents on the grounds that they, and they alone, have special properties that enable them to bridge otherwise unintelligible gaps between various phenomena. Nevertheless, it is important to realize that the felt need to solve the disunity problem only address a philosophically motivated need. The desire to close otherwise unintelligible gaps between mental phenomena is not a concern generated by scientific considerations.

As we have seen, the driving concern that motivates the integrationist story is that "without a cognitive theory the problem [of integration] ... *cannot be solved*. The gap between neurobiology and psychology will be *unbridgeable*" (Gerrans 2014, 36). By this reasoning, we are told that "there *must be* an explanatory relationship between neuroscience and folk psychology" (Gerrans 2014, 33, emphasis added). These are very strong, "musty" philosophical claims—and they are not self-evidently true.[16]

Another clear sign that we are in the realm of philosophy is the ironic fact that the integrationist solution on offer in effect takes the form of a nonnaturalistic autonomy thesis—which is precisely the sort of thing that cognitive theorists strove so fervently to get away from. For in effect the solution to the intelligible integration problem proffered by cognitive theory posits a new space of reasons (a space of reasons Mark II): an autonomous domain of communicating subpersonal and personal-level cognitive phenomena. Cognitive phenomena are set apart from the rest of nature because they don't just brutally interact but intelligibly integrate and interrelate by trafficking in informational and representational contents. Such a theory is allegedly needed because cognitive interactions have features that cannot be understood in merely brutal, causal terms.

Yet the claim that interactions between cognitive phenomena are special in this way mirrors the sort of a priori, analytic demands that proponents of the autonomy thesis insist on. The only important difference is that for cognitivists the illuminating relations between the cognitive phenomena are now to be understood, not in terms of personal-level rational relations, but in terms of the flow and processing of subpersonal information and representational contents.

An age-old adage warns us that we should be careful what we wish for. Likewise, we should be careful with our philosophical demands. Naturalists should be especially wary of demanding that positing mental representations is necessary in the sciences of the mind in order to account for the intelligible character of cognitive integration, for doing so risks turning back the clock in our thinking about causation. As Campbell (2008, 201) warns:

> We naturally seek a certain kind of intelligibility in nature; we naturally try to find explanations that will show the world to conform to reason, to behave as it ought. Hume's point is that there are no such intelligible connections to be found. This point has generally been accepted by philosophers thinking about causation. ... Hume's comments nonetheless do leave us in an uncomfortable position, because we do tend to look for explanations that make the phenomena intelligible to reason. We are prone to relapse, to think that after all we must be able to find intelligibility in the world. ... The lesson from Hume is that there is no more to causation than arbitrary connections between independent variables of cause and effect. We have to resist the demand for intelligibility.

In sum, the worry is that seeking intelligibility in nature can lead us into pre-Humean thinking. Yet beyond this concern, if cognitive theory is to make good on providing genuinely bridge-building and gap-closing unity, it will require more than just talking about how information flow and representational contents are communicated and traded. It requires providing theoretically well-grounded answers of the sort we currently lack about how basic cognitive states of mind can be contentful.

Perceiving

Specifically, in the end, it requires providing a straight solution to the HPC.

Gerrans (2014, 30) is right to observe that "what needs to be explained here is not just the causal interactions among neurones but *the way those interactions* enable cognitive processes and experiences."[17] That is so. But REC supplies just that sort of account by emphasizing the importance of a history of interactions, not by assuming the existence of contentful representations with mysterious properties.

It is a fundamental mistake to think that there can be no cognitive interface between contentless perceiving and content-involving attitudes unless it is mediated by some kind of meaningful communication. RECers hold that when it comes to cognitive integration, what is needed is an *all-action* account, not just, as Mac Davis and Billy Strang famously put it, "a little less conversation, a little more action."

Basic Perceiving Meets Content

For all that's been said, readers might still wonder how REC can possibly explain the integration of contentless and contentful forms of perceiving. For even if it is allowed that REC can explain intra- and intermodal integration when all of the cognition in question is contentless, aren't there other unresolved puzzles? How can purely contentless perceiving systematically give rise to and interface with content-involving perceptual judgments? And how can contentful perceptual judgments influence basic perceiving and how does such perceiving unfold? In general, the question is: How can contentless and contentful forms of perceiving enter into systematic commerce with one another if not—*per impossible*, according to REC—through some kind of contentful communication? The conviction that this question needs answering and that REC cannot answer it would, unless challenged, be a reason to endorse the unrestricted-CIC thesis that all cognition is contentful. The suppressed assumption is that cognitive integration and interaction at every level must always take a communicative form.

REC's answer is straightforward: the interface between perception and thought can be understood in basically the same way as intermodal interaction: with reference to an organism's interactional history.

Interaction and history explain what, why, and how we perceive. We perceive, in Clark's (2015b, 5) helpful formulation, "the patterns that matter for the interactions that matter." And the explanation why we tend to currently perceive such patterns as we do is that we, or our forebears, have a history of engaging in interactions with those selfsame patterns.

Consider the familiar case in which the figure of a dog can emerge from what initially appears to be nothing but a picture of black dots randomly placed on canvas (see figure 7.1).

For those unfamiliar with the picture, at first nothing coherent is seen. But if one is directed how and where to look, given hints where to search, or even explicitly told what to look for, eventually the figure of the dog pops out. When that happens the arrangement of dots look like some*thing meaningful* —they appear as objects, not a collection of random things. Thus it will no longer be possible to see the picture without seeing the figure; one will irresistibly perceive the Dalmatian in the scene.

Importantly, an organism does not need the concept of the thing seen in order for such basic perceiving—which other enactivists call sense making—to occur. However, for those who have

Perceiving 173

Figure 7.1
Camouflaged Dalmatian dog. Reprinted from Lindsay and Norman 1977, with permission of the authors and publisher.

also mastered the relevant sociocultural practices, they will not only effortlessly and irresistibly see a dog, they will be tempted, irresistibly, to judge "It's a dog"—to have a thought with some such predicates and truth conditional content. Here again the individual's history of interactions and engagements, though now bolstered by those of sociocultural kind, is what accounts for the tight connection between contentless and content-involving perceiving.

Notoriously, relations between contentless and content-involving perceiving are not always this smooth. Sometimes the links between basic perceiving and contentful perceptual judgments come apart, as, for example, when most of us encounter the Müller-Lyer illusion (see figure 7.2).

The basic perceptual tendencies of those under the sway of the illusion prompt them to think and say the opposite of what they believe, know, and judge to be true.

This is a clear case in which "a being can interact with its environment as a judger ... on the one hand, and as perceiver and responder, on the other" (Muller 2014, 177). The pull in different directions reflects a tension between two histories of interaction over different time scales: a longer one of basic perceiving, and a shorter one that has additionally involved the

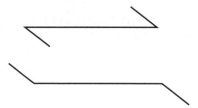

Figure 7.2
A version of the Müller-Lyer illusion

mastery of sociocultural practices of communication, measurement, and other means by which we learn to talk about and to attempt to arrive at the truth about things. Though those endeavors are aligned most of the time, cases of perceptual illusions show they can also come apart.

Capacities for basic perceiving are attuned to what holds over "the long term" in "the majority of situations" (Lupyan and Clark 2015, 282). This leads to the prediction that susceptibility to certain illusions depends on the character of a perceiver's historical environment. This prediction seems to be borne out by findings based on cross-cultural experimental projects using Müller-Lyer stimuli conducted by Segall, Campbell, and Herskovits (1966) on various Western and non-Western subjects, including small-scale societies such as those of South African bushmen, Suku tribespeople, and Bete tribespeople. The results reportedly show "substantial differences among these social groups in their susceptibility to the illusion. American adults in Evanston, Illinois are the most susceptible. … At the other end of the 'susceptibility spectrum,' hunter-gatherers from the Kalahari Desert are virtually immune to the 'illusion.' (They probably would not even recognize it as an illusion)" (McCauley and Henrich 2006, 15).

The carpentered-environment hypothesis put forth by Segall, Campbell, and Herskovits (1966) is one REC-friendly explanation of these findings (for further discussion see Deregowski 1989; McCauley and Henrich 2006). It proposes that the reason some Africans are less subject to the illusion is that they live in environments in which few artifacts have the straight lines and edges of the Müller-Lyer illusion (e.g., their houses and windows are round). Hence their tendencies in basic perceiving are shaped over the long term in ways that differ importantly from those prevalent in the industrialized West.

In sum, REC's duplex account of contentless and content-involving perceiving strikes just the right balance for thinking about our different ways and tendencies in perceiving. It promotes no mysteries about how basic minds come into communion with content-involving attitudes, or about how these can interrelate and integrate without actually communicating. This has the advantage of keeping contentless perceiving and contentful perceptual judgments in their appropriate places for telling the larger story about perceiving.

8 Imagining

The world of reality has its limits; the world of imagination is boundless.
—Jean-Jacques Rousseau, *Emile, or On Education*, Book II

Beyond REC's Reach

As the preceding chapter shows, it is now very much a live question whether perception can be best understood in REC terms. Indeed some theorists already concede that REC might be right about this important portion of the cognitive pie— namely, that REC might be right about the contentless character of basic cases of perceiving. Despite movement on this front, many still incline to the view that REC has inherent limitations and that some version of CIC is needed to understand cases of cognition in which what is thought about is not present—as in paradigmatic instances of dreaming, imagining, planning, and deliberating.

Making precisely this assessment, Clowes and Mendonça (2016) predict a pluralistic future for cognitive science in which it needs to divide up its labors. They hold that REC approaches are well suited for understanding those forms of cognition in which what is thought about is immediately accessible, whereas CIC is needed for understanding forms of cognition that require dealing with items of thought that are less immediate, wholly absent, or nonexistent.[1]

Employing the questionable and ill-understood online/offline distinction, their motivating thought could be formulated in the following way: while basic online perceiving might be content-free, representational content is needed for understanding offline cognition. If the assumptions that motivate this proposal are true, it offers a neat and principled way of carving up the task of understanding cognition: some types of cognition, like perception, are best dealt with by REC, while other types that involve neural reuse and decoupling, such as imagination, are best dealt with by CIC.

The mere fact that this proposal is so intuitively attractive ought to make us wary of violating Ramsey's Rule, as discussed in chapter 1. There is an easy argument for the conclusion that certain forms of cognition must be understood in unrestricted-CIC terms. It employs a simple logic: contentful representations are necessary for thinking about things that are absent or nonexistent—things that are not, or cannot be, objects of immediate perception. Why? Because thinking about what is absent *just is* to represent contentfully. To think of the presence of something in its absence is just what it means to represent something.

Watertight arguments like this come at a price. No substantive grounds are supplied for doubting REC's capacity to understand such allegedly representation-hungry cognition: its ability to do so is ruled out by stipulation. Deploying the presence-in-absence criterion as a means of securing the truth of CIC about these forms of cognition is yet another move from the analytic

Imagining 179

philosopher's playbook. To argue, from the armchair, that cognition about what is not immediately perceivable entails the existence of representational content does exactly nothing to show why positing representational content is necessary for substantively understanding such cognition.

There is a naturalistically respectable way to defend Clowes and Mendonça's (2016) insight about where to draw the line. It is to demonstrate that representational content features in the best accounts of cognition when it comes to explaining how we manage to think about presence-in-absence cases. As a step towards establishing this conclusion there have been attempts to argue for the negative proposition that an enactive approach to mental imagery "is unworkable unless it makes appeal to representations" (Foglia and Grush 2011, 36).

Yet to make fully good on such anti-REC claims requires specifying precisely what kind of substantive explanatory role representational contents per se are supposed to play in understanding mental imagery. Such a justification, it might be thought, already exists in the form of Grush's (2004) emulator theory of representation.

Grush's theory posits the existence of devices called emulators—devices that provide feedback to and help control a given cognitive system by implementing input-output functions that are the same or very similar to those of the controlled system (see Grush 2004, 378–379). REC would be shown to be out of the explanatory running if it turns out both that emulators are part of the best explanation of mental imagery and the detailed account of how they work shows that they involve the manipulation of representational contents.

On the face of it, it looks as if the second condition will be met. Cannonical versions of emulator theory take it that emulators involve the production of efference copies of motor commands and that the efference copies in question "inherit their representational contents from the motor commands that they are copies of" (Mandik 2005, 293; see also Grush 2004, 377).

Interestingly, if a secure case could be made that representational contents do in fact feature in the best explanation of mental imagery then things might be even worse for REC; such a result might well show that REC is entirely false, across the board—and not merely, as Clowes and Mendonça (2016) suppose, that it is unable to shed light on some types of offline cognition. Things would be sweepingly bad for REC if a content-manipulating version of emulator theory carries the day and "perception, including visual perception, results from such [emulator] models being used to form expectations of, and to interpret, sensory input" (Grush 2004, 377).

The idea that perception and imagination are interwoven has become popular in the wake of growing interest in the predictive processing proposal about how the brain does its primary work. PPC paints an image of perception that assumes

> a rather *deep connection* between perception and the potential for self-generated forms of mental imagery. ... Probabilistic generative model based systems that can learn to visually perceive a cat (say) are, ipso facto, systems that can deploy a top-down cascade to bring about many of the activity patterns that would ensue in the visual presence of an actual cat. Such systems thus display ... *a deep duality of perception and imagination*. The *same duality is highlighted* by Grush (2004) *in the "emulator theory of representation,"* a rich and detailed treatment that shares a number of key features with the predictive processing story (Clark 2013b, 198, emphasis added).

REC will not be threatened if all perceiving turns out to be imaginative; not even if the most basic kinds of perceiving are always infused with imaginings. REC would, however, be in trouble if it turns out that representational contents feature in the best account of such phenomena. Indeed, it would be game over for REC if basic instances of perceiving turn out to be fundamentally content-involving because they centrally involve imaginings that are contentful.[2] Hence, when it comes to deciding how best to account for mental imagery, the stakes for REC are pretty much all or nothing.

In this light, it may be no accident that Shapiro (2014a) issued a pivotal challenge to REC with respect to its lack of an account of basic imagining. He deems REC unfit to provide such an account because, in shunning representational contents, he holds it lacks the resources for understanding how we manipulate mental images to solve cognitive tasks such as the Tower of Hanoi puzzles. Naturally, such an explanation will be out of REC's reach so long as it is assumed that this type of problem solving requires, as Shapiro puts it, having "thoughts with the content 'small disc,' or 'medium disc,' or 'large disc.'" (p. 215). Shapiro continues, "I do not need to manipulate the actual discs, because I have representations that, at least momentarily, serve as well (or almost as well). Hutto and Myin do not tell us how these stories might get told without appealing to content" (p. 215).

Answering Shapiro's challenge requires telling a story about mental imagery and basic forms of imagining in REC terms, thus providing an explanatorily adequate account of basic imagining without content. However, before taking up the challenge, it is important to be clear about the properties in dispute.

Many theories of the imagination make use of the notion of representational content and understand it in exactly the restricted sense that is the focus of the REC-CIC debate. Nichols

(2006), for example, tells us that cognitive scientists and philosophers of mind agree on at least two fundamental claims about the nature of the imagination. Crucially, they take it to have representational content: "To believe that p is to have a 'belief' representation with the content *p*. Analogously, to imagine that Macbeth is ambitious is to have an imaginational representation with the content Macbeth is ambitious" (p. 8). Second, they take the capacity for imagining representational content to be "a basic part of human psychology" (p. 9).

REC opposes unrestricted-CIC accounts of the imagination that assume all imaginings, even the most basic ones, always possess contents with some kind of correctness condition, whether such correctness conditions are understood in truth conditional or semantically less demanding terms.

Notably, some accounts of basic imagining attitudes actively reject the idea that they belong to the class of propositional attitudes. Consider Gerrans's (2014) claim that imaginings are cognitively interesting precisely because, even though they are influential and we often act on them, nonetheless they must be distinguished from canonical propositional attitudes, such as belief. Gerrans (2014, 18) holds this precisely because imaginings lack the key semantic properties of the propositional attitudes. On this score Gerrans (2014, 105) tells us that "imagination uses the mind's cognitive resources, such as perceptual, doxastic and emotional processing to create simulations. It thus inherits the intentional structure of these counterpart processes. However *qua* simulations *imaginative states do not have congruence conditions.*"[3]

While REC agrees with Gerrans's (2014) position that basic imaginings lack congruence and correctness conditions, it also allows that some imaginings do involve propositional content. REC's stance on the imagination is, once again, complex and dual-stranded. It does not deny that *some* forms of imagining possess representational content; it denies that *all* imaginings do. Specifically, REC denies that the most basic and primitive varieties of imagination have representational content.[4] The appropriate focus for the debate between REC and unrestricted-CIC on basic imaginings is not therefore a debate about whether any imaginings have representational content, but whether all sensory imaginings—such as imaginings that involve the formation of mental images—have representational contents.

To be meaningful, this debate needs to focus on when and where representational contents actually play a substantial role in accounting for how imaginings do their important cognitive work of enabling creative feats of planning; practicing and executing perceptual-motor tasks; producing works of art; developing novel technologies; and so on.

Trouble in Mind! Imagine That

Defenders of CIC accounts about basic imaginings need to establish that the putative representational contents of such imaginings play a substantive explanatory role in our cognitive lives. This will be an uphill struggle because it is not even clear how such imaginings could have content. Langland-Hassan (2015, 2) neatly sums up the situation: "Much of what has been said about sensory imagination conflicts with the idea that imaginings have substantive correctness (or veridicality, or accuracy) conditions at all." If this analysis is sound, then defenders of CIC accounts of basic imaginings are already on the back foot when it comes to understanding this fundamentally important cognitive phenomenon. Indeed, it looks as if the two standard positive proposals for understanding the representational content and correctness conditions of sensory imaginings—"as-present" and "as-possible" views—are complete nonstarters. What are these views and what problems do they face?

The as-present view holds that the content and correctness conditions of sensory imaginings derive from perceptions and are essentially the same as they would be for perception. Accordingly, imaginings are like perceivings in that they strive to tell us how things presently stand with the world.

This view about the content of imaginings is encouraged by the thought that imaginings just are perception-like experiences that occur in the absence of the things imagined. It may be further encouraged by the fact that basic sensory imaginings are closely related to perceivings in a number of important ways. Not only do both exhibit a similar phenomenology, but it has been discovered that the substantial parts of the same neural pathways are active in both perceiving and imagining. Although perceiving and imagining have different inputs and outputs, the core stages of the processing are similar in structure and function in ways that have promoted the idea that imagining is evolutionarily parasitic on perception (Currie 1995; Slotnick, Thompson, and Kosslyn 2005).

These facts and findings lend initial plausibility to the simulation-of-perception hypothesis of imagining, which holds that basic imaginings are, or centrally involve, perceptual reenactments (Currie and Ravenscroft 2003). This simulation theory of recreative imagining is attractive because it holds out hope of explaining why imaginings are in many ways similar to perceivings and yet differ from them in others (e.g., vivacity).

The best explanation of these facts may well be that imaginings only simulate perceivings but do not replicate them exactly. This hypothesis is empirically credible because of the considerable, but still only partial, overlap in neural processing paths exploited by both perceiving and imagining. The simulationist idea gains support from the general finding that the brain often reuses its neural apparatus to do various distinct kinds of cognitive work (Anderson 2010, 2014).

Clark (2016) cites further supporting evidence that tight connections exist between perceivings and imaginings in their neural underpinnings. One outcome from the imagery-based brain-reading experiments by Reddy, Tsuchiya, and Serre (2010, 97), discussed in chapter 3, is that they reveal that tasks requiring subjects to deal with actual visual percepts and their merely imagined counterparts are "not simply sharing coarse neural resources, but are sharing the fine-grained use of those resources too."

None of these observations entail that perceivings and basic sensory imaginings have essentially the same kind of content and correctness conditions. Indeed, all of the above could be true even if perceivings don't have any kind of content and correctness conditions at all.

This all is good news for fans of the simulation hypothesis about basic imaginings, since there are serious problems with the idea that sensory imaginings carry content that is only correct (true, accurate, or veridical) when imagined properties and objects are in fact present and bring about the imaginings in the right kind of way.

What's the problem? The key issue is that what is imagined is rarely, if ever, present and causing such imaginings in the right kind of way. That being the case, if adopted, the as-present view of imaginative content would have the disastrous result that almost all imaginings almost always misrepresent. Defenders of the view are thus left in "the awkward position of positing a useful cognitive faculty that continually issues in misrepresentations" (Langland-Hassan 2015, 8). When it comes to thinking about the explanatory power of sensory imaginings, it seems the right thing to conclude from this is that they "have an important role to play in successfully guiding behavior. But it is not the same role as perceptual experience" (Langland-Hassan 2015, 6).

The as-possible view fares no better. It maintains that imaginings only represent various scenarios as possible, not as present or even actual (Yablo 1993).[5] There are a number of senses in which the imagination might represent things and situations as being possible. Situations might be construed as being nomologically, metaphysically, or merely logically possible. But for most of the tasks that we rely on basic imaginings to perform, none of these senses of possibility will do. The root problem is that if the content of basic imaginings is construed as only answerable to what is possible, in any of the above senses, then the correctness conditions for sensory imaginings are so unconstrained that they turn out to be almost always successful.

Focusing on metaphysical possibility, Langland-Hassan (2015) illustrates the problem by considering what is required for getting it right when one is engaged in the imaginative task of determining whether a certain sofa seen at the shop will actually fit through one's front door back home. When trying to make that determination one is not interested in nomological, metaphysical, or logical possibilities. Thus, "Suppose that Joe imagines the new couch he ordered fitting through his front door. When it arrives, it does not fit. The couch will have to go back. Perhaps it is metaphysically possible that the couch would fit through the door. Still, it's going back to the store. The imagining was a failure. Our conception of its correctness conditions should reflect that fact" (Langland-Hassan 2015, 9).

This simple thought experiment reveals that the as-possible view of imaginative content and its correctness conditions is seriously out of synch with the kinds of work that imaginings need to do in daily life. The type of content the as-possible view assumes imaginings to have is not suited to the purpose of guiding the relevant cognitive activity.

What we want from imaginings is for them to provide actionable guides as to what is likely to be the case with respect to specific scenarios. In being far too open, the as-possible view suffers from the opposite affliction than that which plagues the as-present view. For on the as-possible view, imaginings are almost always correct, and hence their putative contentful properties are quite disconnected from quotidian uses to which they are typically put. The as-present view has a contrasting problem: imaginings are almost always false despite being in general very useful. Noting these explanatory gaps, Langland-Hassan gives a stark but astute assessment:

On either view, the correctness conditions of imaginings do not track the things they ought to track—things like the helpfulness of an imagining to guiding one's action and achieving one's goals. The successes and failures of imagination, on these views, are *not substantively linked to the cognitive work that imaginings actually do*. The result is that the content and correctness conditions attributed to imaginings are divorced from the functional role they play in the broader cognitive economy. (Langland-Hassan 2015, 2, emphasis added)

In light of the failure of both the as-present and as-possible views, one option is to give up on the idea that imaginings represent anything at all.

This claim that the imagination is contentless across the board has some prominent supporters. Searle (1983), for example, argues that imaginings, as a class, lack a direction of fit. Accordingly, imaginings might be understood as silent about how things stand with the world; they are—to borrow McGinn's (2004, 21) phrase—neutral about reality. However, there are other possibilities to explore before going down the extreme path of denying content to all imaginings.

A Hybrid, Pluralist Solution: Two Takes

There is another, more fruitful way to understand the content and correctness conditions of our imaginings, one sensitive to the specific tasks imaginings help us perform. Langland-Hassan (2015) encourages us to go pluralist and hybrid in our account of imaginings. His leading idea is that imaginings can be entertained in various contexts for a variety of intended purposes. A combined package of content and attitude determines the distinct correctness conditions for any particular act of imagining. The fundamental mistake of the standard theories examined in the previous section is that they try to understand imaginings as a homogeneous class "when conceiving of their content, attitude, and satisfaction conditions" (Langland-Hassan 2015, 11).

Things look more promising if we assume basic sensory imaginings are enlisted to play many different kinds of cognitive roles depending on the surrounding attitude that imaginers adopt toward them. When basic imaginings are integrated with contentful attitudes, they form part of hybrid states that have "both imagistic and non-imagistic components" (Langland-Hassan 2015, 3). Specifically, the first component will be some kind of sensory image and the other, quite distinct, component is "transferred from one's intentions" (Langland-Hassan 2015, 12). When combined, these components form a single imaginative attitude—with its own specific content, correctness conditions, and cognitive role. Crucially, the proposal is that with "both components in play it is possible to understand how imaginings have non-trivial correctness conditions" (Langland-Hassan 2015, 16).[6]

Langland-Hassan (2015) provides some choice examples of hybrid imaginative attitudes, focusing on just two main types, which he dubs judgment imaginings (JIGs) and episodic memory imaginings (EMIs). His purpose is to show how his pluralist solution is meant to work, making clear that the same tools can be used easily to understand a plethora of other imaginative attitudes.

He gives two examples of JIGs. JIGs are imaginative attitudes in which sensory images are put to work in the context of specific judgments. To illustrate, Langland-Hassan (2015) imagines the plight of Avery, who, having never been to Paris, has been misled by images of the Arc de Triomphe into picturing it as silver in color. For this reason, when she adopts a judging attitude toward a sensory image of the monument (as denoted by the words in CAPS in the following content clause) she forms the JIG that: The Arc de Triomphe is A BIG SILVER ARCH. This, as the well traveled know, is false. But importantly, it is only the combined content and Avery's intended use of the image that makes her imaginative attitude false, and indeed something that could be true or false.

Moreover, we can easily suppose that Avery might harbor a quite different intention and might engage in another sort of imaginative exercise even while directing her attention at the very same image. She might, for example, try to imagine what the Arc de Triomphe would look like if it were silver, where she is, this time, under no illusion about its actual color. In such a case, the compound content of her JIG (The Arc de Triomphe painted silver would be A BIG SILVER ARCH) hits its mark; this JIG is true.

Other sorts of complex imaginative attitudes are possible. Langland-Hassan (2015) gives another example of an EMI. Avery might adopt a retrospective attitude on her journey home after her first visit to the Arc de Triomphe. She might endeavor to recall how the monument actually looked when she encountered it on the Champs-Élysées. Let us assume that the combined content of her EMI is that: The Arc de Triomphe was A LARGE WHITE ARCH.

Assume that history favors her (e.g., that Avery has actually been to Paris and seen the correct monument, and so on) and that the image she forms captures the features of its target well enough. In that case we can say she has remembered successfully, and so her EMI is true.

There is much to admire about Langland-Hassan's (2015) positive story. Compared to the manifest shortcomings of its rivals, his hybrid account seems to provide an adequate account of the different kinds of content and correctness conditions that different imaginative attitudes have.

The point, and beauty, of positing hybrid imaginative attitudes is that "we need not double our work by trying to answer these questions about representation in new ways for imagistic states" (Langland-Hassan 2015, 17). The two-component story is an attractive and appropriately flexible way of understanding how sensory imaginings can have any correctness conditions. Crucially, the account is perfectly in step with the specific roles imaginings seem to play in a diverse range of everyday cognitive tasks.

Although Langland-Hassan's (2015) hybrid account is structurally sound and well motivated in these respects, there are reasons to question whether the sensory image component of imaginative attitudes needs to be, or should be, thought of as contributing any kind of content to composite imaginative attitudes. Indeed, the very reasons that Langland-Hassan supplies for wanting to go hybrid about sensory imaginings in the first place should make us wary of this aspect of his proposal.

As we have seen, there are deep problems in providing a general account of the putative content of pure sensory imaginings understood in isolation from surrounding, discursive attitudes and intentions that support and scaffold such images. The lack of any such general account of how imaginings have content ought to make us cautious about simply accepting the idea that a pure, basic sensory imagining, taken as a proper part of a more complex imagining, makes an isolated *contentful* contribution to that whole.

Langland-Hassan (2015) is well aware that any contentful contribution pure sensory images might make would be quite limited. He admits that it is difficult to explain how, on its own, a pure sensory image can be "of" or "about" or "refer" to particulars (other than perhaps its causal source), or how a pure sensory image can be about the past, present, or future. For these reasons, he holds that the contents of pure sensory images should be thought of as being like indefinite descriptions. This is also why the same image can be used in so many different ways in hybrid imaginings (see Langland-Hassan 2015, 16–17).

Indeed, it is because pure sensory imaginings are known to be limited in just these ways that it is necessary to go hybrid.

The proposed hybrid solution works "by pushing some of the burden of explanation onto non-imagistic, discursive thought. ... Specifically, the involvement of non-imagistic contents is important to explaining the ability of imaginings to be about different particulars, to represent counterfactual scenarios, and to be about the past, present, or future" (Langland-Hassan 2015, 17).

Still, despite all of these limitations, Langland-Hassan (2015) presupposes that even in isolation sensory imaginings must have and contribute some kind of content, however indeterminate and ambiguous it may be. Why suppose this?[7] Why not assume that the discursive component does all of the work in making the hybrid attitude contentful? After all, there is a long tradition of assuming that sensory images only have resemblance properties of a kind that are neither necessary nor sufficient for content (see Fodor 1987, chap. 4).

Those considerations are not knockdown. After all, it could be argued that images do carry content after all if, for example, sensory images are just the vehicles for contents that have an independent source (for a discussion see Nichols 2006, 2). That is surely a possibility, but there is no compelling reason to believe such a story in the absence of a credible (and ideally robustly naturalistic) account of how basic sensory images get this putative content.[8]

More to the point, there seems no reason to provide such an account. For even if basic sensory images lack content, this would be no bar to their playing precisely the sort of roles in imagination that Langland-Hassan (2015) identifies.

We can put a REC twist on Langland-Hassan's story if we assume that contentful attitudes scaffold contentless basic sensory images. This REC adjustment to the hybrid story ought not

Imagining

to threaten structural collapse. All that it requires is that it is possible for discursive components to frame noncontentful sensory images such that the resulting product of such a union is a single complex imaginative attitude which possesses content and correctness conditions.

Basic Imaginings at Work: When REC Met MET

To complete REC's account of basic imaginings, what is needed are details of the kind of work basic sensory imaginings, those that entirely lack representational content, might do and how they might do it. Considering what might be minimally needed to explain hominin toolmaking capacities is a good way to think about the kind of work contentless sensory imaginings might do even in isolation from contentful attitudes.[9]

It is well known that the hominins of the Middle Paleolithic had truly impressive toolmaking abilities that would have required considerable cognitive sophistication. For example, fashioning the Levallois flake—the pinnacle of stone-age toolmaking—required the special selection and careful crafting of stone flakes that would fit within other stones to form composite artifacts.

The production of such stone flakes demanded working with sensitive materials, which if mishandled would split and fragment, becoming useless. Worse, such materials were not readily available; they had to be collected from distant sites, requiring toolmakers to engage in advance planning and to have good memories.

It is plausible that basic sensory imaginings played a part in enabling hominins to engage in the kind of mental rehearsals needed for collecting and manipulating materials in their toolmaking industries (see Hutto 2008, chap. 11). Basic sensory imaginings would have made it possible, for example, to remember, and thus find, items that looked like *this*, and to fashion flakes by carefully knapping a stone *this* way, not *that*. Crucially, imaginative capacities are likely to have been important in enabling toolmakers to practice, rehearse, and refine the perceptual-motor manual skills needed for the creation of such specialized artifacts.

What makes the case of hominin toolmaking a useful focus for our purposes is the additional fact that even if hominins enjoyed some kind of primitive protolanguage, they would have lacked anything remotely like the sophisticated practices involving public representations that are a staple feature of modern human societies. In particular, our hominin forebears could not have framed their cognitive activities in the sorts of explicitly discursive ways that we can, thanks to our facility with natural language.[10]

Even so, the cognitive demands placed on both us and earlier hominins by such tasks are not likely to be significantly different. Consider the plight of today's cognitive archeologists and anthropologists who try to learn to fashion tools in the same way that prehistoric hominins originally did in their attempt to get into the minds of our forebears. Mastering the toolmaking craft in this way requires learning how to make such stone tools without relying on anything like publicly accessible diagrams, templates, or linguistically articulated sets of explicit instructions, all of which hominins lacked. Nevertheless, modern humans can learn to fashion such tools without such supports, and in doing so we can reasonably assume that they are employing many of the same basic imaginative capacities that early hominins would have brought to bear on such tasks.

It surely seems possible to fully explain the kind of cognitive activity involved in such tasks in REC terms without positing the manipulation of content-bearing mental representations. Making much of enactivist ideas, Material Engagement Theory, or MET for short, sees "early stone tools as enactive cognitive prostheses capable of transforming and extending the cognitive architecture of our hominin ancestors" (Malafouris 2013, 164).

MET agrees with REC in placing all of the weight on situated, embodied interactions as driving the cognitive activity needed for toolmaking. Thus, "The force and angle of knapping are parts of a continuous process and thoroughly temporal web of interactions that 'involve' complex feedback between limbs, objects, the visual sub-system, and the acoustic sub-system" (Malafouris 2013, 176).

Crucially, MET, like REC, reverses the familiar cognitivist order of explanation: "Stone tools are not an *accomplishment* of the hominin brain, they are an *opportunity* for the hominin brain—that is an opportunity for active material engagement" (Malafouris 2013, 169). Understood in interactive as opposed to representationalist terms, the intentional attitudes that are part of toolmaking practice do not so much stand outside and frame the activity as emerge from within it:

> The directed action of stone knapping *does not simply execute* but rather *brings forth* the knapper's intention. The decisions about where to place the next blow and how much force to use are not taken by the knapper in isolation, they are not even processed internally. The flaking intention is constituted, at least partially, by the stone itself. *Information about the stone is not internally represented and processed by the brain to form the representational content.* (Malafouris 2013, 173–174, emphasis added)

In these sorts of cases intentions are nothing like contentful guides to action that can be teased apart from and imagined to direct embodied activity; rather they are inseparable from and found within such activity. "Decisions" and "intentions" of the basic cognitive variety emerge through interacting and engaging with the material over time.[11] Thus "the locus of early human thoughts stays *with* the body rather than *within* the body" (Malafouris 2013, 174). Here it is important to remember that "the hand is not isolated from the brain" (Tallis 2003, 34). But equally, we must not downplay the fact that hands are not isolated from the objects they manipulate.

Even if this much of the REC-cum-MET story is accepted, it might be wondered if this can be the whole story. As Malafouris (2013, 176) himself notes, even though deliberate, contentful planning is not involved in such tasks, the activity of toolmaking involves "a great deal of approximation, anticipation, guessing and thus ambiguity about how the material will behave."

At this juncture a concern foreshadowed in the first section of this chapter returns to challenge the REC-cum-MET proposal. The concern is that even the most fully engaged, enactive forms of cognition are always infused with imaginings that are needed to inform anticipatory behavior and to guide, or at least adjust, any intelligent engagements. This brings us full circle. Ultimately, this is why Foglia and Grush (2011, 36) claim "the enactive approach to imagery is unworkable unless it makes appeal to representations."

Importantly, however, their dim assessment of REC's prospects in this domain is more nuanced and specific than just presented. In their complete official statement on this topic they claim that "the enactive approach … is unworkable unless it makes appeal to representations, *understood in a particular way*. Not understood as pictures, to be sure. Or sentences for that matter. But those aren't the only options" (Foglia and Grush 2011, 36, emphasis added). Their preferred option is to understand representations as inner models, where "something, M, is a model of X (for some agent A) if A can use M as a stand-in for X" (p. 42).

Why the need for inner models? Focusing on Shepard and Cooper's (1982) mental rotation tasks, Foglia and Grush (2011) consider a situation in which a person is presented with two shapes and must decide if they are congruent.[12] Might the problem be solved in a REC-friendly way by invoking contentless simulative imaginings?

Foglia and Grush (2011) argue that purely simulation-based accounts of such imaginings—which can be easily given a REC treatment—are explanatorily incomplete. Purely simulation-based proposals about the imaginings run into trouble because in focusing exclusively on how embodied activity might be re-created, they fail to provide an account of surrogates for objects that also need to be manipulated. This won't do since, they argue, the latter also play a central role in the completion of the relevant tasks.

Foglia and Grush (2011) maintain that any adequate account of such basic imaginings minimally requires positing at least two distinct components: one that stands in for the embodied activity in question (e.g., manual grasp and a counterclockwise hand rotation) and one that stands in for the object that the activity is directed toward (e.g., a sheet of paper with the relevant shape printed on it; a stone flake in that shape). Prima facie, their claim is quite plausible: embodied activities and manipulated objects appear to be quite distinct, substitutable components in such basic imaginative tasks. Different tasks might involve the same objects, but different actions and vice versa.

Consequently, it seems likely that in such tasks when basic sensory imaginings are used in the absence of overt models, a surrogate for the object in question needs to be used to help solve the relevant tasks. On the assumption that these surrogates are mental models, it follows that positing the existence of mental models is needed for a complete account of at least some basic acts of sensory imagining.

Let us assume for the sake of argument that this is all true—namely, that the best explanation of how certain problem-solving tasks are completed by means of basic sensory imaginings involves the manipulation of mental models of some sort. What follows for REC?

Notably, Foglia and Grush (2011) concede that "action is still of central importance. What makes the model a representation is *precisely that it can be interacted with in a certain way*. So this theory shares a central commitment with the enactivist camp" (p. 9, emphasis added). Indeed, apparently what "most compellingly suggests that they are representations is that *we can engage with them* ... in a manner analogous to how we engage with the [modeled] object or scene" (Foglia and Grush 2011, 8, emphasis added).

Nevertheless, what makes the account offered by Foglia and Grush (2011) seemingly at odds with REC is their insistence that "in all cases, the crucial factor is that something is *being used as a surrogate*, a model, for something else, *and hence* represents that something else" (p. 9, emphasis added).

Of course, this CIC conclusion only follows if using something as a surrogate for something else counts as using it as a representation. For some, it may be that this just is what it means for something to be a representation, in which case the argument is won trivially. But, as the earlier discussion of Ramsey's Rule forewarns, we should be wary of this sort of definitional move. It is also not at all clear how we would decide whether such a definition is appropriate.

As Rowlands (2015a) observes, when framed in standard, unspecified ways—where terms are not agreed on in advance—there is no practical or theoretical utility to debating whether a given phenomenon with a certain profile (e.g., lawful covariance, biological functionality, simulation, resemblance, and so on) qualifies as representational or as playing a representational role. Such debates will only be well grounded if there is a compelling reason for thinking that any one of the many possible phenomena that might be deemed representational is so in some agreed, proprietary sense.

In any case, our question is not the ill-posed question of whether models, inner or otherwise, count as representations *simpliciter*; our question is whether models possess representational contents that do any interesting work in enabling basic imaginings to do their cognitive jobs.

The question that must occupy us is: What explanatorily relevant properties do mental models have? Are such models best understood as having representational contents, or will it suffice for them to carry out their cognitive functions if they are merely relational in character? Do they automatically exhibit "reference" or, in their basic form, do they only possess "ofness"? Do they "stand for" something, do they only "stand in for" something, or do they merely "stand in relation to" something? Or, we might ask: Do such models have semantic properties? Do they have content? And, if so, what role does any content they may have play in enabling them to function as surrogates?

In sum, even if models play a crucial part in enabling us to understand certain important kinds of basic sensory imaginings, this—by itself—gives us exactly no reason to think that representational contents play any role in understanding or explaining the functionally important properties of basic sensory imaginings.[13]

In sum, it may be that using basic sensory imaginings to complete cognitive tasks sometimes involves engaging with surrogate models that exploit systematic structure-preserving correspondences (e.g., certain resemblances) that hold between features of an imagined item and features of some modeled item.

It is plausible that the use of such surrogates makes it possible to respond intelligently even in the absence of modeled items. Even if all of this forms part of the best explanation of how models might enable the completion of certain cognitive tasks, it is important to recognize that such success would still only depend on: certain correspondences holding between the model, the modeled, and the imaginer's systematically engaging with the model appropriately because such correspondences hold. What does not seem to be required, and no argument has been supplied to suggest otherwise, is that imaginers or their subparts need to "take" such correspondences to hold, or that any conceptual or symbolic attributions are made to that effect.

Contrariwise, by focusing on interactions throughout instead of contentful representations, the REC-MET position provides a cleaner, better way of understanding how basic imaginative capacities, including those that involve imaginary models, are appropriately linked to the world in task-relevant ways. REC-cum-MET predicts that even if imaginers do engage with surrogate models such embodied activity is likely to be strongly constrained by the ways they engage with the kinds of things that are modeled and vice versa. This is why we can expect the pattern of engagements to become more refined as they change dynamically over time through experience and practice. In

Imagining

getting beyond the constraining idea that the main job of cognition is to represent how things are or might be, we can better understand how basic imaginings understood as fundamentally embodied and interactive in character play their part in the development of manual skills such as toolmaking as well as providing us one way to complete cognitive tasks such as solving the Tower of Hanoi puzzle.

There seems to be no naturalistically respectable way to rule out a REC account of the imagination on purely analytic or conceptual grounds. Moreover, when we look closely REC seems to give reasonable answers to the question of the explanatorily important work that imaginings do, both in purely basic and more sophisticated hybrid cases. This is so, even if it turns out that modeling of some kind, and not just simulative reenactment, is required in order to complete certain cognitive tasks by basic imaginative means.

The important feature of the inner models posited by Foglia and Grush (2011) is that imaginers interact with them to complete cognitive tasks. It is not clear, and we have been given no reason to believe, that—even if they exist—such models possess representational content. Worse, it is even less clear what work such content could be doing to enable the appropriate interactions.

Thus even if we allow that mental models in some sense of that notion are important for understanding basic imaginings, there seems no compelling reason to suppose that assuming such models have representational contents and correctness conditions will help us to understand how basic sensory imaginings execute their important cognitive offices.

9 Remembering

Diverse kinds of remembering of one's past characteristically get pulled together in autobiographical narrative remembering—semantic memories, all sorts of experiential memories, traces of thought and imagination, fragments of "flashbulb" memory, almost dreamlike sequences …, and much else besides.

—Peter Goldie, *The Mess Inside*

Memory's Many Kinds

Memory is a mixed bag.[1] It has many forms and faces, and there is good reason to doubt that a common essence lies beneath all its teeming variety.[2] At one end of the spectrum we find kinds of remembering that are purely embodied and enactive. At the other end we find content-laden forms of memory. In between, we find forms of pure episodic remembering that appear to require simulative imagination.

In this chapter we show the advantage of adopting REC's duplex account when it comes to understanding each of these varied forms of memory and how they relate to one another. Our strategy is to start at the bottom, jump to the top, and then work our way back to the middle.

Enactive, Embodied RECollections

Procedural memory is the most ubiquitous kind of remembering, found everywhere in the animal kingdom, which also includes humans. Judging by sheer quantity, it undoubtedly constitutes the great bulk of acts of remembering that occur on this planet. Procedural remembering is in essence remembering how to do something. For most nonhuman animals this is usually a matter of remembering how to attract a mate or how to bring down a particular kind of prey. In the human case, such remembering is deeply shaped by cultural influences. For us, most remembering how is a matter of remembering how to do something that involves the manipulation of human artifacts: on that long list, we find remembering how to ride a bicycle, how to pick up a glass, how to open a door, how to type on a keyboard, and the like.

It is important not to equate remembering how with the exercise of blind habits. Remembering how is not just a matter of replicating or repeating a fixed pattern of response in an exact or crudely mechanical fashion (Bartlett 1932, 201–202). Although in one sense remembering how is simpleminded, it is not so simple as to reduce to the mere reactivation of previously stored patterns of embodied response. Such remembering is always context-sensitive; it requires making customized adjustments to novel circumstances in which a current activity or action unfolds. For example, when I exercise my memory of how to ride a bicycle on a new occasion I always make appropriate adjustments to suit the specifics of the current situation: I may not have ridden *this bicycle* before; or maybe not down *this street* before; or at least not while coping with *this precise degree of wind resistance*, and so on.

Moreover, remembering how to execute certain types of actions belongs in the broader class of enactive and embodied remembering (Sutton and Williamson 2014). The class of enactive, embodied remembering also includes not just the ability to reenact general patterns of response but our sensitivity to particulars: the specific details of individual people, places, or things that allow us to re-identify them, such as the characteristics of a lover's face or the features of a favorite café. Thus, we and other animals not only remember how but also remember who, what, and where in purely embodied ways.

All acts of remembering of these types can be understood in contentless enactive or embodied terms insofar as they require nothing more than reinitiating a familiar pattern of prompted response, albeit with adjustments that are dynamically sensitive to changes in circumstance and context.[3]

Pivotally, for our purposes, embodied acts of remembering are marked by the fact that they do not require representing any specific past happening or happenings, and especially not representing these *as* past happenings. These sorts of memory are usefully characterized as embodied precisely because in an important sense such remembering is a matter of reenactment that does not involve representation (Casey 1987).

Here we find the great divide. Declarative memory—by contrast—is a different sort of animal altogether in that it absolutely requires contentful representation. Compare trying to recall and relate, in words, what is involved in cooking a breakfast as opposed to demonstrating that one remembers how to cook a breakfast by simply exercising the capacity. Compare the latter with remembering that you cooked a breakfast on a particular occasion—say, that you cooked eggs Benedict and smoked salmon for your spouse on Valentine's Day. Or compare the latter with remembering that you generally cook that sort of breakfast for your family nearly every Sunday. Explicit, contentful rememberings might feature in an autobiography, but they can only do so because they are not just embodied rememberings but acts of remembering that involve describing facts or episodes in a contentful fashion.

According to this analysis the important difference between nondeclarative and declarative kinds of memory is not that embodied remembering is a kind of amalgamated response pattern built up from an array of repeated performances as opposed to being concerned with very specific, possibly unique, happenings. Rather the crucial difference between the two is that the former does not involve representing any of these particular past occurrences, whereas explicit representation is a necessity for the latter. Nondeclarative remembering differs starkly and fundamentally from acts of remembering that could feature in autobiographical narratives in that the former does not inherently involve any kind of contentful representation of particular happenings. Embodied or enactive forms of remembering, as exemplified by procedural memory, do "not store representations of external states of the world" (Schacter and Tulving 1994, 26; see also Michaelian 2016, 26ff.).

Narrative Practice and Autobiographical Memory

A tale about nondeclarative memory is far from the whole story about memory. There are also declarative forms of memory with which to contend. It is possible that narrativity might not play a central part in every story we want to tell about declarative memory. For example, it is not likely that pure semantic or factual memory, the capacity to remember isolated facts, need involve narrativity. Still, there is at least one prominent kind of declarative memory that can lay strong claim to being inherently narrative in nature.[4] A well-established tradition in developmental psychology regards autobiographical memory as shot through with narrativity.

The principal architects of the Social Interactionist Theory, or SIT, of autobiographical remembering hold that it is a distinctive kind of memory that requires the development and exercise of socioculturally acquired narrative capacities (Fivush 1991; Fivush and Reese 1992; Fivush 1997; Fivush, Haden, and Reese 1996; Fivush and Nelson 2004; Nelson 1988, 1996, 2003, 2007; Nelson and Fivush 2004).

Advocates of SIT claim that, as a matter of contingent fact, in our world "a specific kind of memory emerges at the end of preschool period" (Nelson 2007, 185). In a nutshell, SIT's big idea can be captured in the following formula: with the emergence of "a new form of social skill" comes the emergence of "a new type of memory" (cf. Hoerl 2007, 622, 624).

According to SIT, the emergence of autobiographical memory is socially and culturally mediated in at least two respects. First, autobiographical memory only "emerges within social interactions that focus on the telling and retelling of significant life events" (Fivush, Habermas, et al. 2011, 322). Second, in acts of autobiographical remembering we actively draw on templates found in cultural artifacts that we encounter in the local environment. Hence such remembering is everywhere "modulated by ... sociocultural models" (Fivush, Habermas, et al. 2011, 322).

Narratives enter the story on both fronts. Defenders of SIT propose that the social interactions responsible for making possible the first appearances of autobiographical memory are of a specific kind: they are necessarily linguistically mediated and narrative in form. Moreover, they hold that the sociocultural models that fuel such interactions derive from ambient narratives. Putting these claims together yields the hypothesis that a special sort of social interaction, one involving narrative practices, makes possible a special kind of cognition, autobiographical memory.

Hoerl (2007, 622) highlights the role that SIT thinks narratives play in making autobiographical memory possible by presenting its two central claims as follows:

1. Narratives are the vehicles for a distinct kind of social-communicative interaction.

2. Narratives provide a distinct kind of cognitive framework or format for remembering events.[5]

With assumptions 1 and 2 in play, it is clear in what sense SIT claims that the appearance of unique kinds of autobiographical memory depends on the mastery of certain narrative practices. Becoming a competent autobiographical narrator is not a built-in talent but an achieved skill. The requisite narrative know-how is hard-won and emerges across various stages of ontogenetic development, fostered by "the narrator's mind-enabling and mind-extending apprenticeship in storytelling" (Herman 2013, 94, 230). Children first gain and sharpen their skills as narrators by consuming and producing narratives, where such narratives can be understood as more or less transient cultural artifacts. Specifically, narratives are representational artifacts—they depict a particular series of possible happenings, whether real or imagined (Currie 2010).

It is well known that children's first attempts at narrating are prelinguistic; they typically occur in acts of pretend play, usually when engaging with others, where the creation of such narratives is "accompanied by—rather than [achieved] solely through—language" (Nelson 2003, 28). Yet while SIT recognizes that pretend play is important to the development of the kinds of narrative skills needed for autobiographical remembering, it also regards the basic narrative skills that pretend play affords as insufficient to enable such remembering. The rudimentary narrative skills gained in pretend play need to be further honed, with the active support of caregivers, if they are to eventually make autobiographical memory possible.

Proponents of SIT hold that mastering basic narrative skills in pretend play is not enough because learning how to reminisce about the past requires dealing with others in mastering narrative practices of a decidedly verbal variety.[6] The claim is that only linguistic narratives provide the right kind of public vehicle for sharing the requisite insights and experiences (Nelson 2003, 29). Which insights?

Critically, defenders of SIT hold that engaging with linguistic narratives is necessary for forming an understanding of oneself that contrasts with, and stands over and against, that of another and others—and hence for being able to represent one's past as belonging to one's own personal history.[7] SIT maintains that social interactions that make use of linguistic narratives as objects of shared attention are the means through which children come to discover the existence of divergent perspectives. Thus verbal narratives are taken to provide the medium for the child's first communion with the idea that there are points of view on things other than his or her own.[8]

The major turning point in developing a capacity for autobiographical memory comes, according to fans of SIT, when children are able to have conversations with adults that are aimed at coconstructing narratives about the past. It is through this particular linguistic medium that "children are being confronted with the fact that their memory is not the same as others" (Fivish and Nelson 2006, 240). The following representative exchange, in which M stands for mother and C for child, highlights key features of the process:

M: What other animals did we see at the circus?
C: A giraffe.
M: No, we didn't see a giraffe at the circus. Who did we see at the circus that looked funny? Remember the rhinoceros?
[Four exchanges focusing on seeing a rhinoceros.]
M: What else did we see at the circus?
C: Um, giraffe.
M: No, we didn't. (Fivish and Nelson 2006, 239, reproduced from Reese, Haden, and Fivush 1993)

This is a clear case in which a caregiver attempts to adjust and coconstruct a child's representation of the past, instilling the norms for giving a correct account of the past. It highlights the ways the child learns what is required for accurately representing the past: "The mother and child negotiate what happened, how they felt about it, and their evaluative perspective on the event as a whole" (Fivush and Nelson 2006, 244; Fivush and Wang 2005).

Clearly, quite a lot changes in the preschool and early school period that puts children in a position to enter into and master the kinds of linguistically mediated conversations apparently needed to scaffold their autobiographical narrative abilities. Such conversations with others are crucial, according to SIT, in forging the capacity to remember the past autobiographically. These narratively based and focused social interactions are what enable children to first learn what making accurate claims about the past requires as well as what it is to be a person with a temporally extended existence that can feature in an autobiographical narrative.

There is strong evidence that the process of learning how to narrate one's personal past is neither straightforward nor easy for children. For example, when children two to four years old are first learning to generate such narratives they tend to appropriate elements of another's life story into the content of their own.[9]

There is also further evidence, which speaks in favor of SIT, that the ways adult caregivers narrate events strongly influence "the richness and narrative organisation of children's memory talk" (Harley and Reese 1999, 1338). Individual differences in maternal expositional style—for example, using rich versus low levels of elaboration—strongly predict how the children of these mothers tend to narrate episodes of their pasts (Fivush and Fromhoff 1988; Fivush, McDermott Sales, and Bohanek 2008).

Putting all of this together, it becomes clear why defenders of SIT hold that it is only after children become accomplished narrators in ways that go significantly beyond their first efforts—say in narrating sequences of action in pretend play and dealing with such sequences in picture books—that they get into a position to "view the self as having a specific experiential history that is different from others and thus a specific personal past and a possible specific future" (Nelson 2003, 30).[10]

In sum, by SIT's lights, children only come to be able to think about their pasts in autobiographical terms toward the end of their preschool years because they have begun to master narrative practices through which they "are exposed to an ever-widening circle of understanding people as temporally extended persons with temporally extended minds" (Fivush and Nelson 2006, 242).

And SIT's story does not end there. It draws on further evidence—mainly from studies on Western populations—to defend the view that narrative skills become further refined during adolescence (see Fivush, Habermas, et al. 2011 for an overview; also see Reese and Fivush 2008). It is during this later period of development that autobiographical narratives become temporally well structured and begin to more fully respect local canons about how to tell the story of one's life across time.[11] These late-maturing autobiographical efforts exhibit a much more stable personal frame of reference and are strongly influenced by the local narrative forms and templates.

Developmental psychologists have been especially attracted to SIT because of its potential to explain the well-documented phenomenon of childhood or infantile amnesia—the fact that most of us are incapable of recalling events in our personal history that occurred before the age of around three years (Schachtel 1947; Eisenberg 1985; Pillemer and White 1989; Fivush, Haden, and Adam 1995; Harley and Reese 1999). Explanations in terms of SIT seek to account for our remarkable inability in this regard by appealling to the fact that "something dramatic changes in memory during the years from two to five that makes remembering specific events from one's past life both feasible and of value for the individual" (Nelson 2007, 185).[12]

That something dramatic happens during this period is hard to deny, but it provokes the questions: What exactly changes? How dramatic is the change? And how precisely should we understand it?

Read conservatively, the evidence accumulated by proponents of SIT might be thought to show, minimally, that the mastery of narrative practices significantly adds to existing cognitive capacities, enhancing—albeit in important respects—memory capacities that are already in place. So construed, social interactive narrative practices, though important as scaffolds, would not be responsible for making autobiographical memory per se possible. Instead, the mastery of narrative practices puts children in a position to enhance and improve their recall of the facts by constructively correcting or filling in the "gaps" in the contents of their memory (Barnier and Sutton 2008, 179).

On a weak reading of SIT, mastery of narrative practices helps children to recover, add to, improve, and cement whatever contentful memories they already have, even if those contents are only at first sketchy or only seen through a glass darkly. In line with this, learning how to narrate the past along with others may help children remember things more accurately or clearly and may help make memories stick and become more durable, but the narrative capacities would not be a necessary condition for having memories with autobiographical content.

On this construal, memories with autobiographical content would be initially formed and stored even though they would be hard to faithfully recover until they are embedded in narratives. Such a weak reading is consistent with the SIT assumption that acquiring a capacity for narrating the past is something individuals can only normally gain by interacting socially with others.

The weaker, enhancement reading of SIT, therefore, sponsors a more modest understanding and explanation of childhood amnesia, according to which "the reason why adults cannot remember any particular events that happened during their early childhood is that their own memory, at the time, was *not geared up* for retaining memories of such particular events" (Hoerl 2007, 625, emphasis added).

Yet SIT is more frequently advanced as a more radical claim—namely, that the mastery of narrative practices makes possible a wholly new and unprecedented kind of memory (Nelson 2007, 184; Fivush, Habermas, et al. 2011, 322). Unlike the enhancement view, the strong reading of SIT holds that prior to gaining the capacity to construct meaningful narratives about the past there is no possibility of representing the past autobiographically at all. Fivush and Nelson (2006) advocate this stronger reading when they write:

> Until children begin talking with others about what they have experienced in the past, or about the experiences of others at some other time, young children *do not represent themselves in past lives or project themselves into possible futures*. ... Parent-guided reminiscing, specifically about internal states, structures children's developing understanding of their own and others' past and allows for [makes possible] children's construction of a temporally extended understanding of self and other. (p. 235, emphasis added; Fivush 2001)

Pressing for the strong reading of SIT, Fivush and Nelson (2006) maintain that "talk about the past, in particular, *is essential* for children's developing understanding that memory is a representation of a past event" (p. 240, emphasis added). What this makes clear is that the strong reading of SIT differs from the enhancement reading precisely in claiming that prior to mastering the relevant narrative practices, there is no properly *autobiographical content* to personal memories.

Strong SIT claims that it is the mastery of narrative practices that provides children with the very first opportunity to meaningfully recall the past in a properly autobiographical manner at all. Thus, if strong SIT turns out to be true, it follows that there is no preexisting autobiographical content to our memories to be enhanced prior to our learning to narrate the past along with others.

The Puzzle of Pure Episodic Remembering

Can SIT account for purely episodic, nonautobiographical kinds of memory? The question is pressing since humans, at least, are apparently capable of having some kind of episodic memories before their mastery of the relevant narrative skills and independently of their exercise of them.

Yet since SIT claims that such skills must be brought to bear if there is to be properly autobiographical remembering, it seems that there are capacities to remember things about our personal past in ways that (1) do not reduce to the kinds of nondeclarative, nonnarrative embodied forms of remembering (described earlier in the chapter) but that nevertheless fall short of (2) sophisticated, narratively based sorts of autobiographical remembering.

The urgent need for proponents of SIT to address this challenge is made evident if we consider Nelson's various formulations of the relations between autobiographical and episodic memory (for a discussion see Hoerl 2007, 623ff.). On the one hand, Nelson speaks as if autobiographical memory were a subtype of episodic memory—one marked off from other kinds of episodic memory in that it makes special reference to the self. For this reason she clearly distinguishes autobiographical memory from pure kinds of episodic memory (see Nelson 2007, 186). This way of formulating the relation between autobiographical and episodic memory is problematic for the strong SIT she advocates since, if true, it would entail that autobiographical memory isn't truly a distinctive kind of memory at all, but only a special subvariety of episodic memory, albeit one that possesses a particular kind of content.

On the other hand, in places Nelson also writes as if the two notions were equivalent, as in the phrase "episodic (and thereby autobiographical) memory" (Nelson 2007, 184), or that the former entails the latter. Equating these notions or regarding episodic memory as necessarily linked with autobiographical memory is problematic for SIT as long as it is accepted that episodic memory is present early on in development, prior to the development of narrative practices. For in that case equating or strongly linking the two notions would entail that references to the self would have to be possible prior to developing the relevant socially scaffolded capacities that are needed to narrate the past. That would render SIT, in either version, manifestly false. Or it would require, implausibly, holding that "episodic memory is not available in the early years of life" (Nelson 2007, 187).

Thus clarifying the claims of SIT and evaluating their prospects of being true requires getting clear about how episodic memory and autobiographical memory might be related.

On one, helpfully neutral, construal episodic memory "refers, roughly, to the form of memory responsible for allowing us to revisit specific episodes or events from the personal past" (Michaelian 2016, 5). Episodic memory is often taken to be a form of declarative memory that involves representing the past in explicit and consciously accessible ways (Michaelian 2016, 27). However, a virtue of the neutral formulation of episodic memory supplied above is that it does not require that all such memory is declarative in these respects. Nor, importantly—for reasons that will become clear shortly—does it demand that episodic memory involve any explicit representation of the self.

How then might we understand pure episodic remembering? A number of theorists have proposed that episodic memory is simply a form of recreative or simulative imagining that enables us to construct and entertain possible episodes (Gerrans and Kennett 2010; De Brigard 2014; Michaelian 2016).[13] Inspired by scientific developments around this topic, it has been claimed that there is no intrinsic difference between remembering and imagining. Starkly put, the simulation theory of episodic memory holds that "to remember, it turns out, is just to imagine the past" (Michaelian 2016, 14, 120).

According to this simulation theory, at its core, episodic remembering is really a matter of generating "self-centered mental simulations about possible events that we think may happen or may have happened to ourselves" (De Brigard 2014, 173).

There is convincing empirical support for the hypothesis that there is a common cognitive basis for acts of memory and imagination (Szpunar, Watson, and McDermott 2007; Schacter, Addis, and Buckner 2007; Schacter and Addis 2009). Novel scientific work on mental time travel has repeatedly confirmed the existence of some strong similarities in the patterns of neural activity associated with the sorts of cognitive procedures employed in thinking about our past and imagining our possible futures.[14]

Acts of recall in which specific events or episodes are reexperienced—for example, when we remember what it was like for us to enter a particular classroom for the first time—can be understood in terms of acts of re-creative imagination that involve neural reuse and reactivation. As noted in chapter 7, the simulationist theory is supported by evidence that the brain often opportunistically reuses its existing neural apparatus for novel tasks (Anderson 2010, 2014).[15]

In line with this simulation theory, De Brigard (2014, 177) hypothesizes that episodic remembering is made possible by a particular operation of a larger cognitive system—an operation that enables us to entertain hypothetical thoughts about possible happenings. Thus he offers an account of episodic memory as "an integral part of a larger system that supports not only thinking of what *was* the case and what potentially *could be* the case, but also what *could have been* the case" (p. 158).

Drawing on insights from PPC, De Brigard proposes that this sort of remembering consists in the optimal reconstruction of a previous experience, where optimal reconstruction is understood as "a retrieval process probabilistically constrained both by schematic knowledge and the frequency of prior encounters with the target memory" (p. 171). He is attracted to this view because of empirical findings that show many ordinary memory distortions present events in ways that are coherent and plausible (p. 173).[16] The claim is that at least when we set our imaginations to the task of remembering the past, we constrain them to generate plausible episodes of what may or may not have happened—possibilities appropriate to the situation in question (p. 179).

Gerrans (2014) identifies the larger system in question with the Default Mode Network (or DMN), and he claims that in such cases the DMN provides imaginative simulations that can serve as narrative elements (p. 13). Indeed, Gerrans suspects that the DMN "evolved to produce narrative fragments" (p. 17).

All of this is consistent with the suggestion that when not constrained by tasks of remembering the past or predicting the future, the DMN drifts into the free play of the imagination witnessed in daydreaming, dreaming, and delusion. Or, in other words, "When not organised for problem solving the DMN reverts to the screensaver mode we experience as daydreaming or mindwandering. In this mode there is no overarching goal to provide narrative structure" (Gerrans 2014, 62).

For our purposes, what is important is that that even in episodic remembering, the imaginative episodes that constitute the narrative elements or fragments of which Gerrans (2014) speaks are "not always assembled into full-scale narratives, they may remain fragmentary and episodic, but their cognitive nature is to be the building block of a story assembled from subjective experiences" (p. 13).

Autobiographical memory proper is something more than pure episodic recall, where the former is characterized as going beyond "recalling the who, what, where, and when of an event, to include memory of how this event occurred as it did, what it means, and why it is important" (Fivush, Habermas, et al. 2011, 322).

One distinctive characteristic of autobiographical memory is the particular way accounts of the self get bound up in the recall of past happenings. Prior to being able to construct autobiographical narratives, insofar as children are capable of a kind of episodic remembering, "There is no distinct self associated with the contents of memory" (Nelson 2007, 186). Even though such a claim may seem counterintuitive to those who assume that all experience necessarily entails having a sense of self, Nelson (2007) explains that in the case of pure episodic remembering, there is no necessity for and no ground for supposing that the self appears as a content of the remembering. This is because "there is no contrast; there are no contents that are independent of the activities and interests of the self, therefore the self is implicit in all, but not 'self-evident'" (p. 186).[17]

All in all, if we endorse the simulation vision of pure episodic memory, it is possible to hold that prior to the development of autobiographical memory, younger children do have memories about particular events, "although their memory of the components of such experiences may be *scattered and fragmented*, reflecting the character of their experience in terms of the limits of short-term memory for connecting segments of extended experience" (Nelson 2007, 186). Claiming this is entirely consistent with holding that in their very early careers children are not capable of "re-experiencing a specific episode in the sense of thinking 'I was there'; rather the data can be explained simply in terms of recalling 'things that happened'" (Nelson 2007, 186).

Combining the simulative pure episodic remembering with SIT's account of autobiographical remembering can explain the fact that "when an adult does remember something from earlier than age three, it is usually a fragment or a brief scene, not a full meaningful episode" (Nelson 2007, 185). Given all of these considerations, we can understand the significance of the discovery that we "talk about the shared past with partners, friends, family, or colleagues in order to facilitate or tap what may be only fragmentary, partial, or shrouded in our own memories" (Barnier and Sutton 2008, 179).

Allying strong SIT's take on autobiographical memory with a simulationist theory of episodic memory has the further virtue of opening up the possibility of developing an even more fundamental defense of the idea that autobiographical remembering depends on scaffolded social interactions. For it might be argued that it is only through the process of mastering linguistically mediated practices that one encounters the kind of cognitive friction needed to learn how to make contentful claims, and thus for getting things right or wrong—about the past, or any other topic—at all. If so, mastery of special sociocultural practices may be required, for a more fundamental reason than merely to enable children to form representations about their personal past that have properly autobiographical content. More pivotally, mastery of such practices might be necessary to enable them to make any truth-evaluable, contentful claims about the past at all.

This thesis about the link between linguistic practice and contentful representation is far from uncontroversial.[18] Yet—if it turns out to be true—it would advance the fortunes of a strong reading of SIT since it would entail that only individuals who acquire the ability to fashion autobiographical narratives about their past could have *any* properly meaningful, contentful thoughts about their memories.

Nor is it necessarily at odds with the assumption that purely episodic remembering is grounded in simulative imaginings. Many theorists who defend a simulation theory of imagining hold that even though simulative imaginings make a bona fide cognitive difference, they are nevertheless unlike familiar cognitive attitudes in that they lack inherent correctness or congruence conditions.[19] Thus even if we assume that episodic remembering is a kind of simulative imagining, there need be no conflict in thinking that making truth-evaluable claims about one's personal past is a capacity that requires a special kind of social immersion and support.

Roles and Functions of Remembering

Even if the puzzles concerning nonautobiographical episodic remembering can be adequately addressed, a strong reading of SIT may still strike many as an intuitively implausible thesis about what is involved in remembering our personal past. This is because it is at odds with the prevalent contention that the basic function of remembering is primarily to extend our representational capacities beyond perception. According to a familiar picture, perception informs us about how things stand with the world via the senses.

Our perceptual systems' takes on the world are deemed contentful in that they form representations that can be true or false, or more or less accurate. Memory, so the picture tells us, extends our cognitive contact with the world beyond the here and now of perception. This is achieved, according to a modern variant of this story, by retaining and manipulating information that when recovered and recombined in the right ways enables individuals to reproduce their contentful takes on past happenings. If all goes well, and the reproduced representations are faithful, we remember; if not, we misremember.

The beating heart of this view, designated the content-based approach, is the assumption that the function of memory is "to store, retain, and then to reproduce (or make available, or reconstruct) the contents of past experiences" (De Brigard 2014, 158).[20]

On the flipside, a strong reading of SIT gains credibility when considered against the backdrop of new thinking about the function of autobiographical memory that is empirically motivated. Although the content-based approach remains a mainstay of orthodox analytic philosophy of mind, its vision of the core function of autobiographical remembering has been put under enormous pressure by a wealth of empirical findings accumulated over the past thirty years.

Combined scientific findings using various paradigms paint a picture that reveals the great extent to which our memories are easily susceptible to outside influences—influences that can drive them to distortion and error without our noticing (Brainerd and Reyna 2005).

In an effort to cast doubt on the content-based approach and to make room for new thinking about the role and functions of remembering in our lives, De Brigard (2014) catalogs a wealth of empirical work that outlines the many ways outside influences ordinarily distort and bring about false memories. His strategy is clear. Step 1: Establish that if we assume that memory is, at root, all about the faithful representation of past happenings, then we must be prepared to confront and accept the overwhelming evidence of "pervasiveness of misremembering" (De Brigard 2014, 156). Step 2: Expose the theoretical difficulties that the content-based view encounters in trying to maintain that the primary function of remembering is faithful reproduction of past content in light of the fact that it everywhere so spectacularly and reliably fails at that task. Taken together, the empirical findings and failings of theory add up to an "important challenge for the philosophy of psychology and cognitive science … to reassess our understanding of the function of episodic autobiographical memory" (De Brigard 2014, 157).

It is pivotal to De Brigard's argument that, on the content-based view, memories are distorted to the extent that they are not entirely faithful reproductions. For, assuming this, distortions creep into everyday acts of remembering whenever we switch between *field* and *observer* perspectives, exhibit the *telescoping effect*, or *extend the boundaries* of remembered scenes, as we regularly do.

Consider a common form of perspective shift, between field and observer perspectives, that often occurs when we remember. To remember events from the *field* perspective is to experience them from the point of view from which we originally experienced them (Nigro and Neisser 1983). If autobiographical memory is meant to be the straightforward recovery or replay of our past experiences, then we would expect all personal memories to be experienced in that way.

Yet we often recall autobiographical events from a perspective other than the perspective from which we first experienced them. That is, when remembering we sometimes experience events in our personal lives as a sort of witness or spectator—in other words, as an *observer*—of those happenings. Indeed, it is not unusual for us to switch between field and observer perspectives when remembering.[21]

We are also subject to the *telescoping effect* when remembering the past such that the temporal ordering of events gets confused. This effect usually kicks in after a three-year window, and results in recent events being perceived as more distant than they are and distant events being, vice versa, perceived as having occurred more recently than they did (Neter and Waksberg 1964; Thompson et al. 1996; Janssen, Chessa, and Murre 2006).

Completing the list of everyday memory distortions, we also typically undergo *boundary extension* when remembering. This is a distortion of commission—one in which more of a scene is remembered than was experienced, typically in ways that plausibly anticipate how things would have looked just beyond the edges of what was originally seen (Intraub and Richardson 1989).

Beyond the fact that our memories are regularly subject to such distortions, we are also remarkably prone to false memories. These are easy to invoke. It has been shown time and again that "people are prone to misremember perceptual details of previously witnessed events, and even entire events that never happened in their lives" (De Brigard 2014, 161).

The misinformation paradigm in psychology reveals the great extent to which memories can be significantly influenced by postevent misinformation. In a classic, groundbreaking experiment Loftus, Miller, and Burns (1978) showed subjects slides of a car stopping at a stop sign. Then they manipulated some of the descriptions of what one group of subjects saw—stating that the sign was in fact a yield sign. It was found that subjects in this test group were more likely to report seeing a yield sign than those in the control group who were not misinformed.[22]

Similar results have been frequently replicated in other experimental formats. For example, Loftus and Pickrell (1995) showed that false memories can be implanted by getting people to read short narratives about plausible childhood events that never occurred. Through this technique it was famously discovered that up to a quarter of subjects could be induced to falsely remember having been lost in a shopping mall when they were children despite such events never having happened (De Brigard 2014, 161).

Wade et al. (2002) and Lindsay et al. (2004) showed how robust this tendency is by achieving similar results using faked photographs of a family trip in a hot air balloon. They succeeded in getting roughly 50% of test subjects to report having experienced this event in their childhood, again, even though this was not the case.

Other laboratory research highlights our liability to an *imagination inflation* effect, whereby merely being asked to imagine what it would have been like to experience a particular event can lead people to mistake having actually experienced such events in their personal pasts (Garry et al. 1996).[23]

All of these experimental findings have been further bolstered by the Deese-Roediger-McDermott paradigm, which makes use of semantically and perceptually associated lures planted in lists of items to be memorized—for example, asking if the word *sleep* appeared in a list of items associated with sleeping. The results show that participants were prone to report having seen the nonpresented lures 55% of the time (Roediger and McDermott 1995).

Nor are these tendencies restricted to laboratory settings. Notoriously, Loftus (1996) extended this experimental work to more ecological settings, exposing the ways eyewitness memory can be chronically inaccurate about the details of allegedly experienced events (see Loftus 2005 for an extensive overview of this research). In a similar vein, French, Garry, and Mori (2008) showed that romantic partners tend to rely on each other, more than do strangers, to support acts of remembering. However, it was also shown that this procedure tends to make romantic partners complicit in each other's memory errors rather than more accurate.

What do all these empirical findings imply for our theories of memory? De Brigard (2014) holds that if we understand memory through the lens of the content-based view, then we are problematically forced to conclude that (1) misremembering is a common phenomenon and (2) we don't even notice whether we are misremembering.[24] In a nutshell, he tells us, "We usually go about our lives without questioning the accuracy of our memories—many of which are, probably, inaccurate" (De Brigard 2014, 162).

Certainly, in light of the accumulated evidence it seems clear that "autobiographical memory does not reproduce specific past events with precision" (Campbell 2006, 363).

Fans of the content-based view, as De Brigard (2014) notes, can try to accommodate this fact by availing themselves of a Millikan-style defense about proper functions (Millikan 1984). That is, it is open for them to argue that memory may have the proper biological function of providing faithful representations of past events even if it rarely produces such representations. This would be akin to noting that sperm are meant to fertilize ova, even though—statistically speaking—the great majority of individual sperm reliably fail to fulfill this function, as discussed in chapter 5. Although appealing to proper functions to try to rescue the content-based view is technically possible, it is hard to defend the hypothesis that faithful remembering succeeds often enough, and in the right circumstances, so as to make its hypothesized representational work as biologically beneficial to us as sperm are to the continuation of species.

To make this proposal plausible it would have to be shown that statistically rare instances of faithful remembering can reliably confer some very important biological benefit. Those who wish to defend such a claim need to identify what such rare instances of faithful remembering might do for us. Making that case won't be easy in light of the fact that we have no biologically basic means of reliably telling whether our memories are getting things right or wrong in any given circumstance. Therefore, a better answer for supporters of the content-based view is to argue that even though memory aims to produce faithful representations, the reason this does not happen more frequently is because of predictable temporal and other environmental constraints that limit the encoding of information. If that were so, it would explain why we initially only encode "relatively sketchy or 'gist-like representations' of previous experiences" (De Brigard 2014, 166). And this in turn would explain why, in many ordinary acts of remembering, we are forced to reconstruct our memories in ways that result in imperfect representations.

Accordingly, the fact that we do not typically encode and store all of the available information about particular events need not lead us to doubt that the function of memory is to produce faithful representations, even if it usually fails to achieve this result.

However, as De Brigard (2014) observes, this proposal is problematic because it predicts that we should be less prone to distortion and error in remembering under conditions that favor more complete and faithful information encoding and recovery. But this does not fit with

> many of the experimental results ... in which false and distorted memories occurred in informationally poor environments where the stakes are pretty low—paradigmatically in psychology labs in which only short movies or brief word lists are presented. The amount of information participants experience in these environments is significantly lower than the amount of information one normally experiences in every-day situations. (De Brigard 2014, 166)

In this light, rethinking the main function, or functions, of autobiographical remembering as something other than faithful representational reproduction becomes attractive. These considerations all point to the conclusion that "it is a mistake to think of memory as a system that is uniquely—or even primarily—dedicated to reproducing the contents of previous experiences" (De Brigard 2014, 177).

Looking afresh at what autobiographical memory might be doing for us is preferable to supposing that it reliably fails to perform a biologically basic function that, given how we know it actually operates, is "almost impossible to achieve" (Campbell 2006, 365).[25]

What alternative function or functions might autobiographical memory be performing for us? There are grounds for thinking that such remembering plays important roles in establishing social cohesion and that it is bound up primarily with our use of self-narratives. Such narratives help us regulate expectations, plan for the future, and steer actions, both collective and individual. Shared remembering, such as in the retelling of family stories, typically provides something "besides or beyond accuracy" (Barnier and Sutton 2008, 179). In such contexts, remembering together has less to do with representing the past accurately and more to do with forging and maintaining "intimate and longstanding relationships" (Barnier and Sutton 2008, 181).

Engaging in narrative practices, those that involve giving accounts in rich storied content, has been shown to correlate positively with mental health. A number of findings demonstrate that people "who are able to narrate the emotional events of their lives in more self-reflective ways show better physical and psychological health" (Fivush, Bohanek, and Zaman 2010, 46; see also Fivush et al. 2003). In particular, it has been found that

individuals whose narratives include more causal explanatory language (words such as *because*, *thus*, and *understand*) and more emotional language (the inclusion of both positive and negative emotion words) subsequently show lower anxiety, lower depression, higher sense of well-being, and higher immune system functioning than individuals who use less of this kind of language. (Fivush and McDermott-Sales 2006, 126)

Other research reveals that choice of narrative, but not necessarily its accuracy, is important to our well-being, showing that "how we remember the stressful events of our lives has an impact on our ability to cope" (Fivush and McDermott-Sales 2006, 125; McDermott Sales et al. 2005). This sort of coping is not a matter of recovering details of past happenings accurately so much as a forward-looking basis for dealing better with other, similar stressful future happenings.

These observations chime well with Campbell's (2006, 372) claim that autobiographical remembering may be an activity that aims at shaping "the possibilities for how we go on." Indeed, overemphasis on the idea that memory is fundamentally about representational fidelity, as promoted by the content-based view, has surely contributed to little attention having been paid to the affective dimensions and social benefits of autobiographical memory and remembering together.

Yet even if we abandon the one-sided picture of the *primary* function of memory being that of accurately representing past happenings, none of these observations should lead us to suppose that autobiographical remembering never involves attempting to get at and establish the truth about what happened in the past faithfully.

Getting claims about the past right is undoubtedly one of the central things we hope to achieve in our acts of autobiographical remembering. Nor is there any essential conflict between our attempts to seek the truth about the past via autobiographical remembering and the fact that such remembering takes a narrative form. For some narratives aim at conveying truths.[26]

Taking everything into account, the lesson we should derive from the sum of empirical findings to date is that remembering is open to influence from others. We are not entitled to conclude, however, that such influences need always be wholly negative. Remembering with, or under the influence of, others can cut both ways, to produce both "beneficial and troubling outcomes" (Barnier and Sutton 2008, 179). On the one hand, the evidence clearly shows that autobiographical remembering is vulnerable to misinformation, distortion, and corruption. Yet, on the other hand, it is surely possible that reminiscing with the support of others can—at least under the right circumstances—enable us to "successfully renegotiate the emotional significance of a shared past experience, to arrive at both a more accurate picture of the past and a more fruitful conception of current self and other" (Barnier and Sutton 2008, 179).[27]

This rethinking of the functions of autobiographical memory fits with SIT, even in its strongest version. For—as noted at the end of the previous section—it is possible that we are only able to even raise questions about the truth or accuracy of our memories once we have learned the ropes of a particular kind of claim-making narrative practice. If so, it will not just be because until they have mastered the relevant narrative practices children cannot form representations with autobiographical content, but more fundamentally, because until they have mastered such practices children are not in a position to make claims that represent their pasts in ways that can be true or false, full stop.

It may be that getting at the truth of past happenings is not fundamentally unlike getting at the truth of things in any other domain. Seeking the truth about our past is a sophisticated business, one we are only able to conduct at all because of our familiarity with the norms of this peculiar social enterprise. Typically, success in this endeavor requires the care and effort of many people. Yet even when conducted alone, we must call on socially instituted norms and procedures since "one of the demands of recollection is that we are prepared to be critically attentive to the concepts, narratives, feelings, and self-conceptions through which we experience the past" (Campbell 2006, 374).[28] Consequently, just as a strong variant of SIT would have it, there is reason to think that being in a position to make claims about the past, let alone to determine their veracity, is not a solo feat that can be performed by individuals by means of their unculturated, biologically basic cognitive hardware alone.

All told, getting the best, most comprehensive account of memory will require adjusting some long-held philosophical intuitions about its various faces, characters, and functions. By introducing the distinction between contentless and contentful kinds of cognition, REC provides the appropriate set of tools for developing a new positive story about memory. Although telling the full story about memory in REC terms would take a much longer book, we hope to have provided compelling empirical and theoretical reasons for trying to give that fuller account in due course.

Epilogue: Missing Information?

What is outside the head may not necessarily be outside the mind.
—Lambros Malafouris, *How Things Shape the Mind*

Don't Mess with Mr. In-Between!

For all that has been said, readers may still feel that something is missing from REC's account of cognition, something indispensable to any story of cognition—that is, they may find REC's story unsatisfactory because it says nothing about the information that they assume is acquired, processed, pooled, mapped and remapped, and generally made use of by the brain.

Take Michaelian (2016). He agrees with REC in thinking that pure episodic remembering is constructive and should be understood through the lens of simulation theory, but he continues to think the story of such remembering needs to "assign an important role to information storage" (p. 8)

In taking this stand, Michaelian (2016) stresses that he is in direct opposition to the Wittgensteinian views of Moyal-Sharrock (2009) and Stern (1991) about how to understand memory. It is useful to juxtapose Michaelian's opposition to what REC and the Wittgensteinians have to say on this issue with Sutton's (2015) remarks concerning REC's negative agenda in clashing with the mainstream, as discussed in the preface. Doing so brings out precisely how negative critiques and positive contributions to the field can sometimes go hand in hand.

Sutton (2015) reports having been inspired to pursue productive work by Stern's (1991) Wittgensteinian criticisms about the limitations of the storehouse metaphor of memory, even though he disagreed with Stern on some details. For this reason Sutton (2015, 413) holds Stern's approach up as a shining example of how philosophical work can inform science through "engagement rather than conflict or isolation."

However, in the same passage, Sutton draws a sharp contrast in this regard with those Wittgensteinians, such as Moyal-Sharrock (2009), who have fallen under the sway of REC. Sutton (2015, 414) maintains that where they "go astray is partly in their claims about what psychologists and cognitive scientists do and believe, and partly in their choice of criticisms, or their sense of which issues matter most." But, as we are about to see, it is on the very issue of adjusting thinking on "what matters most" to the mainstream—in particular on adjusting its commitment to the idea that information is processed and represented by the brain—where some of the vitally important theoretical work remains to be done.

Returning to Michaelian (2016), his case is interesting precisely because he agrees with REC on so much, yet balks when it comes to making the transition to the revolutionary side of the street by letting go of his commitment to the information processing paradigm. He holds on, in the end, to the idea that acts of remembering depend on various manipulations of originally stored information.[1]

But is an account of how the brain represents and uses information in fact needed in order to give a simulative account of cognition in terms of recreative reenactment? Why assume stored information or contents play any part in simulatively reenacting an experience if stored information and contents are not needed for procedural memory? After all, Michaelian (2016) accepts with respect to procedural memory that there is nothing to declare and nothing that is declared. On these issues he fully agrees with RECers in holding that "reference to storage of rule-based information in non-declarative memory is redundant, since an appeal to changes at the neural level is sufficient to explain the relevant changes in the organism's behavior. ... The assumption that rule-based information is stored does no additional explanatory work" (p. 26).

Indeed, Michaelian (2016, 26) stresses that "we need not mention information processing at all in order to give a complete description of nondeclarative memory." This is because, on his analysis, while "nondeclarative memory does involve the modification of the brain of the organism on the basis of its experience ... [it is] ... unlike declarative memory, [because] ... it does not involve the modification of the brain of the organism *as a means of making information available to the organism* in the future" (p. 26, emphasis added).

In contrast, RECers see no reason at all to suppose that declarative memory differs in this key respect from nondeclarative memory. Neither form of memory gets its work done by making information or content stored in the brain available to the organism. Consider the discussion of pure episodic memory in the preceding chapter. Such remembering counts as a kind of declarative memory for Michaelian (2016). Yet it is not at all clear why the capacities for experiential reenactment that feature in pure episodic remembering, even though they recreate partial responses in neural pathways, would need to make use of or convey stored information. This is especially so if we take seriously the arguments of chapter 7 cautioning us to resist the philosophically driven temptation to think that information processing and representations are needed to explain synchronic intra- and intermodal integration—since this integration would surely feature in such simulative efforts to reenact experienced episodes.

Neurodynamics

This raises a pivotal question. If REC is right that pure episodic remembering does not involve information processing, then must we deny that such remembering is cognitive? Apparently, we must if we accept the standard cognitivist criterion for the cognitive—that which takes information processing to be its unique mark. For example, it is because Michaelian (2016, 27) accepts that criterion that he deems nondeclarative remembering as being in "an important sense not a kind of cognition."

RECers think that's a strange verdict for a naturalist to reach: it looks like one motivated solely by a questionable a priori philosophical demand. But if we are correct in thinking that information processing is not involved even in the so-called higher forms of cognition, then following this same cognitivist logic, we would have to draw an even stranger conclusion. We would have to say that all forms of declarative memory—indeed, all we take to be paradigmatically mental—are in fact noncognitive.

To avoid such a peculiar result we prefer to revise our understanding of cognition along the revolutionary lines set out in chapter 1. We explicate further what that change amounts to in the remainder of this closing chapter.

Making a convincing case for an alternative REC vision of neurodynamics requires answering Aizawa's (2015, 761–762) pressing questions:

Presumably, the brain has something to contribute to the production of … "cognition," but what is this? Perhaps the brain processes information, manipulates symbols, or transforms representations. If it does that, then very little indeed has been changed.

On the other hand, maybe the brain does not contribute information processing or symbol manipulation or the transformation of representations. But, if not, then what does it do?

REC's suggestion is that cognition should be understood, quite generally, in terms of processes "of creating differences that make a difference, that exist [in part] only because of what the brain does" (de-Wit et al. 2016, 10). But what does the brain do? We have provided reasons for thinking neural activations are not actually "coding," "computing," "conveying," or "communicating" information. Information is not *really* picked up, or passed on, or pooled in the brain for use by the brain.

Neurodynamics, understood à la REC, takes the form of informationally sensitive, well-connected neural activity that plays influencing and mediating, as opposed to representational, roles in enabling organisms—and here we reiterate our favorite line from Clark (2015b, 5)—to "get a grip on the patterns that matter for the interactions that matter." In line with REC's endorsement of teleosemiotics, such well-calibrated neural activity systematically influences organismic responding by maintaining connections with specific worldly features in "good" cases, but it does so without representing anything or being represented by anything in the brain. Importantly, we don't need a replacement notion of information, other than content-less covariance, to understand the correspondences to which the brain is sensitive.[2]

It helps to compare REC's content-free account of neural dynamics with Cao's (2012) full-fledged biosemantic take on alleged communicative activity within the brain. Cao investigates the extent to which conceiving of parts of the brain as sending and receiving contentful signals has any warrant or explanatory payoff. She considers a range of candidate neural signals (action potentials, neurotransmitters, action potentials across many neurons) and receivers (whole pre-synaptic cells, whole post-synaptic cells, assemblies of neurons). In the end, she concludes that while areas of the brain may be trafficking in semantic information of some minimal variety it is really only when we look at whole organisms that we "see the both flexibility and competence to act, together giving rise to a richer semantic story" (Cao 2012, 67). Her final verdict is that, "the real epitome of a semantic information-using system is not the brain or parts of it, but the whole organism" (p. 70).

Whether focusing on individual cells, neurons or whole neural populations as possible content consumers, in each case Cao (2012) holds that such neural structures could only possibly receive "vanishingly little" content (p. 65). For example, although she allows that the post-synaptic density acts in ways that make it credible to count it as a receiver of semantic contents, she nevertheless acknowledges that, "the world of this sender receiver system would be claustrophobically small. It would be representing something like "I'm active! I'm active!" (p. 65). Summing up, she tells us that:

It seems more reasonable to ... give up the common interpretation that a single neuron is doing something like representing 'a very small piece of the world outside the organism' though indeed its activity may be well correlated with the structure of that small piece of the world. Instead, the role of the single neuron is more like that of the man inside Searle's Chinese Room—taking inputs and systematically producing outputs in total ignorance of their meaning and of the world outside. The single cell responds to local regularities and rewards, which as a result of evolution have become coordinated (in some complicated fashion) with external regularities and distal rewards for the whole organism. (p. 66)

RECers agree with Cao's analysis, as quoted above, verbatim—providing "evolution" is understood broadly so as to include individual development and learning. We hold that what she concludes about single neurons applies to activity in the brain across the board. We strongly diverge from Cao (2012), given our reasons for wanting to steer clear of the official versions of biosemantics as set out in chapter 5, in that we see no basis for sticking with the idea that brains or their parts transmit or receive any content, not even of a minimal variety.

How, in adopting its contentless vision of neurodynamics, does the REC take on the work of the brain square with some of the most important discoveries in the field—such as the Nobel Prize–winning finding about place cells, their brethren, and the brain's positioning system (O'Keefe and Dostrovsky 1971; O'Keefe 1976; O'Keefe and Nadel 1978; Hafting et al. 2005; Moser and Moser 2008)?

We have already discussed, in chapter 2, the ways REC is and isn't compatible with Kandel's (2001, 2009) explanations of how learning and memory are linked. On the one hand, REC does not question the details of Kandel's explanations of how experience leaves a trail of structural changes in the brain. On the other hand, it challenges the standard cognitivist interpretation of what that Nobel Prize–winning work really achieved. The story is exactly the same with respect to REC's proposed treatment of these other Nobel Prize–winning discoveries about the existence of place cells, grid cells, and head-direction cells in the hippocampus of rats.

Place cells fire when an animal occupies a specific location within its environment. Such cells act like a kind of GPS system that can be exploited for various tasks, when supported by other cells with diverse firing conditions—such as head-direction and grid cells. The exciting discovery was that a pattern of cell firings corresponding to specific spatial locations enabled rats to explore untraveled paths toward previously seen targets. Rats, it was found, are able to explore novel navigational possibilities offline, while dreaming (Ólafsdóttir et al. 2015).

Does this groundbreaking research falsify REC? Does it show that rats or their brains are using inner cartographic maps or models composed of contentful representations? Many will assume so. But we must tread carefully here.

First of all, there is a particular risk of confusion on this score due to "the profusion of things that can be called 'models'" (Godfrey-Smith 2009, 33). Also, in light of the discussions in chapters 5 and 7, we must be especially careful not to confuse evidence that a system has map-like or model-like properties for evidence that the system is actually using a contentful map or model.

Godfrey-Smith (2009, 33) tells us that in modeling "the state of one thing X is consulted in dealing with another Y. One thing is treated as a guide to the other." Something similar is true of maps. Assuming this, we can ask whether the brain literally consults the positioning system like a map, model, or guide to the navigational possibilities of the environment.

That would be to picture the brain as a homunculus, leading to well-known trouble. It would be better to say that the positioning system has properties that are map-like and those properties are exploited in extremely complex ways by various neural activities, enabling the organism to respond to the relevant navigational possibilities. There is a significant theoretical difference between these two interpretations, and REC is untroubled by the second.

The real issue between REC and its unrestricted CIC rivals turns on the question of whether evidence concerning the positioning system entails that information is represented in and used by the brain in a contentful way. Bechtel (2016) supplies an impressively detailed analysis of the positioning system research with just this question in mind. He concerns himself solely with the question of whether the research under discussion is committed to assuming that "the brain represents information in ways that can guide behavior" (p. 1290).

Bechtel's analysis is not just detailed, but his general approach is sophisticated; he completely sets aside the question of whether scientists "are correct in their assumption that there are representations in the brain" (p. 1317). Instead he concerns himself wholly with whether and to what extent these scientists are invested in and committed to the idea that they are dealing with a representational system. His concern is with what the scientists take their task to be, and what they think their work aims and is directed at.

After much meticulous analysis, Bechtel concludes that the relevant neuroscience research takes itself to be occupied with "determining which neural processes are content bearers and understanding how they represent what they do" (p. 1291). The conclusion he draws from his investigation is that, "Content characterizations are not glosses on the research; the goal of the research is to determine what content the representations have" (p. 1317).

Bechtel allows that like all research programs, this one "may eventually stall and be replaced by an alternative that rejects understanding brain activity in representational terms" (p. 1289). But, of course, as things stand the research in question is still flourishing, and in any case Bechtel presses the point that if it is said that these scientists are not "fully committed" to the idea that brain activity really serves a representational function, then "it is not clear what their research has established" (p. 1316). Moreover, he claims that "independent of that objective, there would be no reason to conduct [such] research" (p. 1288).

We agree with Bechtel (2016) that there is a sense in which if representational talk were nothing but a mere gloss, it would be difficult if not impossible to make sense of the commitments that have been driving these seemingly flourishing research programs. He makes a compelling case that it is only because these scientists conceive of their quarry in representational terms that they have been able to get such a seemingly powerful and impressive grip on what appear to be the working parts of an important mechanism of cognition.

But that is not the end of the story. For even if the representational talk in question is not a mere gloss it might be a translucent sheen. So even if it is not entirely opaque it may still introduce significant diffusion or distortion. Conceiving of their goals in CIC terms might be helping to keep scientists on the right track even if, conceptually speaking, their understanding of what they are keeping track of is less than fully transparent.

The theoretical disagreement between REC and CIC is a refined disagreement in which small details matter. Getting clear about such details is always tricky.

What then is the crucial difference between these rivals? It amounts to this: It is one thing, as REC holds, for information—understood only in terms of contentless correspondences—to be used by a system in various ways that guide an animal's behavior in various tasks. It is quite another thing, as unrestricted CIC holds, for information—understood only in terms of contentless correspondences—to be contentfully represented within and consulted by the system to generate an animal's behavior in various tasks.

Neuroscientists engaged in the sort of research Bechtel (2016) so painstakingly describes are likely to be unaware of the subtleties of the philosophical debates surrounding content, such as those examined in chapter 5. In particular, they are likely to be unaware of the nuanced philosophical candidates for understanding contents or intentional objects. Note what Bechtel (2016) writes: "The only variable that *was correlated* with firing rate was location and so this was what *the researchers treated the cells as representing*. Discovering that the firing of a particular class of neurons depended on the rat occupying a particular place field was sufficient" (p. 1296, emphasis added).

The point is that even though the scientists in question do talk about representation, that fact alone does not tell how to understand the commitments of such talk—as discussed in Rowlands (2015a) and the preface with reference to Huw Price's polysemic use of the term "representation." Whether or not the scientists use the term "representation" in their theorizing should not distract us from the really important questions. After all, we are not interested in the use of labels but in which properties the scientists need to commit to in their theorizing and which properties show up in the best account of the relevant phenomena. The theoretical debate turns on how, after careful analysis, we should best understand and explicate the nature of the cognition.

In conducting some kinds of scientific work, attending to these subtle differences may not matter. But these devilish details surely do matter if we are to get a cleaner, leaner, and clearer understanding of the properties of cognition. Here, we are reminded by the quotation from Feynman, Leighton, and Sands (1963, 1–2) that features in the initial epigraph to this book, "Even a very small effect sometimes requires profound changes in our ideas."

Importantly, the approach we adopt sees philosophical work as necessary for understanding the true theoretical commitments in this domain. Whether our proposals are accurate or inaccurate, philosophers engaged in this sort of work are not mere onlookers who aim to report only what "the science" says. Philosophers can play an active part in helping us understand the phenomena of interest. We offer a REC rendering of neurodynamics with the aim of clarifying the conceptual framework that cognitive and neuroscientists are operating with, making it possible to avoid what would otherwise be deep and perhaps intractable theoretical mysteries.

In any case, without an answer to the HPC, if we stick with the unrestricted CIC rendering, what can we or the scientists really say about the nature of the positioning system findings? How should we understand and explain their allegedly contentful character?

By REC's lights, the core focus of future research in neuroscience should continue to be to determine how, and in which ways, brain activity enables organisms to respond sensitively to contentless information—for example, whether this is by means of dynamic, distributed waves or static patterns of activity in particular areas, or some combination.

Neuroscience ought to continue to focus "on the development of methods and techniques that enable us to study what causes what in the brain" (de-Wit et al. 2016, 12). REC offers a theoretically sophisticated and parsimonious way of making sense of such work without introducing unnecessary theoretical extravagances and allows us to avoid deep theoretical mysteries.

Extensive Dynamics

Still, doesn't this answer create a different problem for REC? If we assume REC's contentless-correspondence interpretation of the world-relating neurodynamics of the positioning system, doesn't the mere existence of neural mechanisms argue against REC's claim that cognitive phenomena are fundamentally extensive?

Not really. Patterns of place cell firings get used in the wider system to produce novel, appropriate navigational behaviors when embedded in the right environmental contexts. The links and connections to those environments are nonaccidental. As Ólafsdóttir et al. (2015, 1) report, "Dominant theories of hippocampal function propose that place cell representations are formed during an animal's first encounter with a novel environment and are subsequently replayed during off-line states to support consolidation and future behaviour."

As Bechtel (2009, 544) says elsewhere of mechanistic explanation, it requires consideration not just of the parts and operations of the mechanism "but also of the organization within the mechanism and the environment in which the mechanism is situated. Accordingly, mechanistic explanation in psychology requires not just looking down (decomposing the mechanism), but also looking around (recomposing the mechanism) and looking up (situating the mechanism)."

REC holds that a characteristic feature of basic cognition is that it is fundamentally world-involving. This claim is motivated by REC's take on Ur-intentionality and teleosemiotics, laid out in chapter 5. For to understand the basic intentional characteristics of cognition is to understand it as constitutively bound up with the worldly objects that are its special concern. Motivated by that vision, REC sees cognitive processes as wide-reaching, spatially and temporally extended forms of embodied activity. Cognition is—at root—extensive in ways that make REC's take on the extent of mind different from versions of extended functionalism that assume cognition is only, occasionally, and in special circumstances, extended.

In embracing teleosemiotics, REC is also motivated to adopt the view that cognition is fundamentally extensive because of its skepticism about the existence of mental contents and about the idea that brains literally represent and process information.

We have identified a host of theoretical problems with representationalism throughout this book: understanding how information could be literally processed by brains; understanding how information could be represented by brains; and understanding how content could be a natural feature of basic cognition. RECers believe that the combined weight of those problems tips the scales in favor of nonrepresentational theories of basic cognition.

All of this together should lead to a rejection of internalism, and a move beyond the extended-mind debate as traditionally formulated. Some doubt this. Objecting to REC's take on the extent of basic cognition and questioning the feasibility of its revolutionary ambitions, Wheeler (2015, 13) has complained that "even if the relationality of cognition, as identified by radical enactivism, is a genuine phenomenon, it is manifested at the wrong level of explanation to deliver extensive externalism."[3]

Wheeler (2015) thinks that REC's explanatory focus is only on the personal level, whereas, he holds, the real cognitive action must be at the so-called subpersonal level since it is only at that level that we find the mechanisms that generate intelligent behavior and responsiveness. In line with the venerable tradition discussed in chapter 1, he believes the search for underlying mechanisms justifies thinking that the subpersonal is the right level at which we find the truly cognitive.

Yet if representationalism goes, and along with it the content/vehicle distinction, so too goes the best prospect for providing a nonarbitrary "mark of the cognitive" at the so-called subpersonal level.[4] In other words, to surrender representationalism, importantly, is to relinquish the only credible, nonarbitrary candidate for a criterion or mark of cognition that cognitive science has served up to date.

There are general reasons to be wary of relying too heavily on the distinction between the so-called personal and subpersonal levels (see Drayson 2014). Certainly, REC's reasons for moving away from representationalism put the standard cognitivist take on that traditional distinction into doubt. But if we are to speak in such terms, then with important caveats, REC need not deny that part of cognitive science's job ought to be to identify subpersonal mechanisms (see Wheeler 2015, 14). Nevertheless, REC cautions against trying to identify any mechanisms that might be discovered to play a part in the larger cooperative enterprise of enabling cognition with cognition itself. Indeed, it warns against doing so even if it should turn out that the mechanisms in question play a central role in such enabling work.

Anyone who hopes to supply a mark of the cognitive at the level of mechanism, without the backing of an information processing or representational theory of cognition, is facing an uphill battle. They must address a serious problem: How to nonarbitrarily demarcate and individuate cognitive states in purely scientific terms, focusing only on what happens in sub-personal terms? To answer this puzzle surely requires having an account of which subpersonal properties confer cognitive status. Representationalism could have supplied such an account.[5] But if representationalism is false, what might do such work in its place? Simply appealing to facts about what goes on neurally or mechanically during acts of cognition won't do.

In his later career, Putnam (1988, 74) came to see the root problem with all such endeavors clearly: "One looks for something definable in nonintentional terms, something isolable by scientific procedures, something one can build a model of. ... And this—the 'mental process'—is just what does not exist."

To illustrate the sort of difficulties bound to be encountered in any such enterprise, consider Rupert's (2009, 2010) attempt. He claims that "if there is any theoretically interesting divide between what is distinctively cognitive and what merely causally contributes to intelligent behaviour, it is to be found in the persisting, integrated nature of cognitive architectures" (Rupert 2010, 344).

Rupert (2009, 45) defends the view that any mechanisms which are distinctively cognitive will be located "inside the organism either entirely or in the main." Importantly, in advancing this claim Rupert (2009) only calls on his minimal systems-based principle—a principle that does not refer to or assume that cognition involves information processing, representations, or content. It is formulated as follows: "[A] state is a cognitive state if and only if it consists in, or is realized by, the activation of one or more mechanisms that are elements of the integrated set [of members] which contribute causally and distinctively to the production of cognitive phenomena" (p. 42).

Let us suppose, for the sake of argument, that we have a precise means of individuating well-integrated mechanisms from the rest of nature. Let us also assume that we can determine precisely which mechanisms are implicated in any token bout of cognitive activity. Let us assume further that we are reliably able to discern which are the more, or even the most, integrated mechanisms that make a distinct contribution to any given bout of cognitive activity.

Even in this best-case scenario, without a supporting mark of the cognitive for backing, it is not clear what would license treating even the most integrated mechanisms that enable some cognitive activity as picking out the cognitive per se. What could justify such a conclusion? Further grounds, other than stipulation, are needed to justify holding that well-integrated mechanisms that make distinctive contributions pick out the truly cognitive.

For Rupert (2009) the requisite justification comes not by calling on a substantive theory of cognition but rather from the alleged empirical fact that finding integrated mechanisms is where all explanatory action resides. He holds that the discovery of well-integrated mechanisms is what best "accounts for the successful practice in cognitive psychology" (p. 43).

It is true that such discoveries are a major part of the explanations in cognitive science—the discovery of the positioning system in rats is a perfect case in point. Yet such special cases of success do not provide a secure basis for the generalization that only well-integrated internal mechanisms can do the explaining in the sciences of the mind. What basis is there for concluding that all of the explanatory successes in the sciences of the mind can be put down to the operations of well-integrated internal mechanisms?

Many cognitive processes are too complex and heterogeneous to make it credible that explanations that only focus on the behavior of integrated mechanisms can account for all that needs explaining in the cognitive sciences. As Silberstein and Chemero (2013, 960) argue, "Systems biology and systems neuroscience contain robust dynamical and mathematical explanations of some phenomena in which the essential explanatory work is not being done by localization and decomposition."

Against that familiar assumption, Silberstein and Chemero hold that the "explanatory work in these models is being done by their graphical/network properties and the dynamics thereupon" (p. 960). They further observe that to dismiss the possibility that focusing on such properties might be truly explanatory would be "as extreme as thinking that implementing mechanisms are irrelevant for explaining cognition and behavior" (Silberstein and Chemero 2013, 960; see also Dale, Dietrich, and Chemero 2009). If these authors are right, then looking for internal mechanisms simply isn't the only effective explanatory strategy for cognitive science.

Silberstein and Chemero (2013) take this thought further, defending the idea that dynamical explanations are nonmechanistic. Other voices around this issue adopt a more conservative view but are equally ardent about not limiting cognitive science to explanations supplied in terms of internal mechanisms however well integrated they may be. For example, Zednik (2011) argues that there is good reason to think that dynamical explanations are needed in cognitive science. He describes what he takes to be two very different kinds of dynamical explanations: Thelen and Smith's (2001) explanation of infant perseverative reaching and Beer's (2003) explanation of perceptual categorization in a minimally cognitive agent.

Zednik (2011) differs from Silberstein and Chemero (2013) in thinking that dynamical explanations are mechanistic and, hence, that the mathematical tools and concepts of dynamical systems theory are, at least in these cases, being "used to describe cognitive mechanisms" (Zednik 2011, 239). He concludes that "dynamical explanations are well suited for describing extended mechanisms whose components are distributed across brain, body, and the environment" (p. 239).[6]

It is still an open and hotly debated question whether our understanding of mechanisms will widen so as to incorporate dynamical phenomena under its umbrella. Still other, more conservative voices who caution against collapsing these notions also recognize the important contributions that dynamical thinking makes to our understanding of cognitive phenomena.

For example, Kaplan (2015, 759), resists the idea of nonmechanical dynamical explanations, holding instead that "dynamical models do not compete with mechanistic models; rather, dynamical models provide one important set of resources, among many other resources, for describing aspects of mechanisms. The relationship between dynamics and mechanism is one of subsumption, not competition." Even so, Kaplan is quick to note, however, that this conclusion is consistent with the possibility that dynamical approaches may offer "a novel framework for conceptualizing cognition and neural activity in nonrepresentational or non-computational terms" (p. 759).

It is not clear at this stage how best to understand the relation between dynamical systems and mechanisms, nor what their precise contributions to explanation and understanding are. This situation should make us especially cautious about attempts, like Rupert's, to identify cognitive phenonomena by appeal to the unique explanatory contributions of integrated mechanisms without the backing of a mark of the cognitive grounded in an independent theory of cognition. Crucially, it seems that if REC is right and representationalism has to go, then without an equally powerful replacement mark of the cognitive we are bereft of the resources needed for nonarbitrarily individuating cognitive wheat from chaff.[7]

This outcome should not be seen as a cause for despair. For the debates concerning how best to understand extensive dynamics or the dynamics of extensive mechanisms reveals that we may have interesting new resources to consider when thinking about cognition as something that happens out in the open rather than as inevitably behind the scenes. Thus, the continuing debates about how to understand the exact contributions and implications of dynamical thinking are encouraging developments in light of REC's claim that if we go nonrepresentational in our approach to cognition, there is no clear scientific rationale for thinking cognitive activity must reside in some internal domain that is smaller or more limited than extensive, world-involving activity.

Loops into Culture

REC aims to free up our thinking about the extent of cognition. But its proposal about the extensive nature of minds is not limited to basic forms of cognition in which organisms are directly coupled to their environments in wide reaching, rolling patterns of perceiving and acting. As we have been at pains to demonstrate throughout this book, REC is also committed to the idea that some minds also loop into society and culture and vice versa. Cognition can and often does reach beyond the individual such that when it comes to understanding a great deal of cognitive activity the right unit of analysis requires focusing on spatially and temporally distributed processes—or to borrow Sutton's (2015) apt phrase, cognitive ecologies—that are bound up with our patterned practices, customs, and institutions. Thus we agree with Clark (2016, 138) in holding, for example, that "the anticipatory activity of embodied, predictive brains is also shaped over slower timescales by being situated in culturally crafted, designer environments of our own making."

Seeing cognition as extensive in this way opens the door to broader lines of research in the cognitive sciences—research that aims to better understand some kinds of cognitive activity in terms of socioculturally distributed, extensive practices and processes—practices and processes that can give rise to and play a major part in explaining some quite special and significant forms of cognition. As we argued in chapter 6, recognizing the powerfully transformative effects of enculturation, the analyses of this book prepare the ground and open the door for further research that seeks to supply "an account of how cultural content and normative practices are built on a foundation of contentless basic mental processes that acquire content through immersive participation of the agent in social practices that regulate joint attention and shared intentionality" (Ramstead, Veissière, and Kirmayer, 2016, 1).

REC's story does not end here. In securing the theoretical foundations for thinking about the most fundamental forms of cognition as contentless, defending the possibility of accounting for the natural origins of content-involving minds without explanatory gaps, and showing how basic minds meet content—in perceiving, imagining, and remembering—we believe we have taken crucially important steps in articulating REC's positive story, thus opening the path for further efforts in evolving enactivism.

Notes

Chapter 1

1. E-approaches have collectively launched many constructive, fertile research programs that focus on an extremely diverse array of cognitive phenomena. Those E-approaches that make central use of the explanatory apparatus of dynamical systems theory have notably provided "a fresh perspective on many foundational problems in cognitive science, including perception-action, memory, word recognition, decision making, learning, problem solving, and language" (Riley and Holden 2012, 593). Big claims have been made about such E-approaches. They are said to be "gradually supplanting" their traditional cognitivist competitors (Cappuccio and Froese 2014, 3; Stewart, Gapenne, and Di Paolo 2010). Some go further, attesting that E-approaches have now "matured and become a viable alternative" to traditional representational-cum-computational approaches to cognition, yielding theoretical and methodological advances that "avoid or successfully address many of the fundamental problems" faced by their rivals (Froese and Ziemke 2009, 466).

2. Brook (2007, 5), who identifies the historical roots of these ideas, reminds us that "Descartes conceived of the materials of thinking as representations in the contemporary sense. And Hobbes was the first to clearly articulate the idea that thinking is operations performed on representations. Here we have two of the dominating ideas of all subsequent cognitivist thought: the mind contains and is a system

for manipulating representations." Chomsky (2007), a principal architect of the most recent cognitive revolution, is also clear about this historical debt in identifying these developments as a "second" cognitive revolution that built on the previous cognitive revolution of the early modern era.

3. We speak here only of sensorimotor enactivism as presented in Noë 2004, and make no claims about Noë's position in his later writings.

4. Hurley and Noë (2003, 342) are emphatic about their commitments on this score: "We propose that perceptual quality can be explained by the ways neural states figure in dynamic sensorimotor patterns. Is this a version of functionalism …? No."

5. In making this important adjustment REC invites questions about cognition that only make sense when it is understood along processual lines, such as: "Why, for example, did the process unfold in this way rather than that? Why has it not stopped? What is sustaining it and keeping it going? What is responsible for any regular patterns we may observe in its progression? And so on" (Steward 2016, 78).

6. According to Noë 2004, content is a feature of perceptual experience. Thus he tells us that "mere sensory stimulation *becomes* experience with world-presenting content *thanks to* the perceiver's possession [and exercise] of sensorimotor skills" (p. 183). And he makes clear that the claim that perception has world-presenting content is interchangeable with the claim that it has representational content. See Hutto 2005 and chapter 2 of Hutto and Myin 2013 for more details on the CECish character of Noë's early views on perception.

7. And, of course, endorsing CIC is possible for E-theorists because "nothing in the enactivist view requires abandonment of contentful states" (Shapiro 2014a, 216).

8. Machery (2009) attributes precisely such a view to Prinz (2002, 113ff.), stating that the latter has proposed that "perceptual representations are whatever psychologists of perception say perception involves" (p. 110).

Chapter 2

1. As Rosenberg (2014b, 27) emphasizes in discussing Kandel's research, "The difference between humans and rats and sea slugs is of course proportionately larger numbers of neurons in more complicated circuits are involved. ... [Yet] what is going on in all three cases is just input-output wiring and rewiring."

2. Memories have long been compared with archived items that can be faithfully retrieved by minds, as if they were the sorts of things that exist in a kind of internal mental storehouse. Down through the ages memories have often been conceived of as images—proxies of items encountered by the senses—that are received, sometimes suitably augmented, retained, and later retrieved by minds. This familiar picture of memories has a long and influential history, finding perhaps its earliest and most eloquent expression in St. Augustine's *Confessions*. Whether St. Augustine meant to take this storehouse analogy as a figurative device or a literal proposal about remembering is a matter for scholars to debate. What is clear is that other thinkers, inspired by the same analogy, have regarded it as a serious basis for philosophical theorizing about mind and knowledge.

3. Friends of cognitivism readily acknowledge that "no philosopher has ever claimed that covariation by itself constitutes or confers content" (Shapiro 2014a, 216; see also Matthen 2014, 124–125; Miłkowski 2015, 78). Yet they fail to recognize the full implications of this admission.

4. See Piccinini 2015, 30, for details of other possible ways of understanding information.

5. In the passage quoted Clark (1997) talks about reasoning, but Clark and Toribio (1994) make it clear that such reasoning is to be understood in a very broad sense. For example, keeping track of and behaviorally anticipating something not present is taken to be an instance of "reasoning about the absent" (see Clark and Toribio 1994, 419).

6. Elsewhere, in Hutto and Myin 2014, we suggest that it amounts to adopting a Totemic Theory of Mind, since totems too are concrete

figures that represent what spiritually binds together otherwise disparate things.

7. Dennett (1991) makes clear that "the general principle of the content/vehicle distinction is relevant to information-processing models of the brain. ... In general, *we must distinguish features of representings from the features represented*" (147, emphasis added).

8. The standout case is, of course, the remarkable range of changes in practice that followed in the wake, over long tracts of time, of the switch to Copernican cosmology from its Ptolemaic predecessor. The changes that flowed from that transition were not restricted to rethinking our theory of the heavens. The impact of the Copernican revolution was clearly felt in religion, philosophy, and literature but in more quotidian domains too, say, in the way it altered our agricultural and navigational practices. Indeed, if Churchland (1979) is to be believed, taking Copernican theory on board—fully and completely—could potentially shape, quite fundamentally, how we perceive and experience the world. These are all examples of the ways in which, "theory is regularly an intimate part and constituting element of people's second-by-second practical lives" (Churchland 1993, 218).

9. In the surrounding discussion Aizawa (2014) hits the nail on the head, again: "One might dissolve the problem or abandon the problem, if one rejects the traditional concept, but one cannot solve it" (p. 40). See Hutto 2013a for further discussion of how deeply held framework assumptions can, at once, generate impossible conceptual problems as well as apparently irresistible explanatory needs.

10. The fictionalist approaches discussed here must not be confused with the anodyne view that talk of content may have practical or instrumental value even if we don't take it seriously in terms of metaphysics. For "one benefit of this talk is that it provides us with a way of referring to the (real) neural causes of behaviour" (Sprevak 2013, 555). Thus even if attribution of content to cognitive states is systematically false, it can still provide a way of labeling mental states "and hence of keeping track of them" (p. 555).

11. Sainsbury (2010) questions the general logic of fictionalism along these very lines. In a nutshell, Sainsbury's complaint is that in trying to make metaphysical savings by assuming that the ontology of some region of discourse is fictional, "No progress towards nominalism has been made. ... Mathematics is a fiction, a story. But what is a story?" (p. 2).

12. McDowell (1998, 349) claims that explanations involving subpersonal contents have an "enormous capacity for illumination" despite being "irreducibly metaphorical." But he fails to make it clear how this can be so (see Hutto 2013c for a critique).

13. Despite making this claim, it is unclear to what extent Matthen (2014) actually endorses an optimistic realism as opposed to some version of antirealism. He vacillates in what he says on this matter. On the one hand, he speaks of a multiplicity of modules that communicate by sending and receiving signals, remarking that "on natural assumptions, these communications have a semantics" (p. 123). On the other hand, he also remarks: "I am not suggesting that this kind of agentive talk should be taken literally. My point is that it provides a design perspective on the machine without which you cannot comprehend the setup" (p. 122).

Chapter 3

1. Elsewhere Clark (2016, xvi) repeats this claim, again insisting that PPC is a "compelling 'cognitive package deal' in which perception, imagination, understanding, reasoning, and action are co-emergent from the whirrings and grindings of the predictive, uncertainty-estimating brain."

2. Clark (2016) is very clear that we must distinguish two stories that might be told about predictive processing. The first presents a general, and as he puts it, "extremely broad," vision according to which the brain role's in cognition is that of "multi-level probabilistic prediction" (p. 10). The second, more specific predictive processing story attempts to give the details about how the first story might be told in light of current developments in neuroscience.

3. Clark (2015a, 3) identifies the radical implications of PPC as follows: (1) the core flow of information is top-down—the forward flow of information is replaced by the forward flow of prediction error; (2) motor control is just top-down sensory prediction; (3) efference copies—the putative copies of motor commands—are replaced by top-down predictions; (4) cost functions are absorbed into predictions. Despite proposing such major changes to the standard cognitivist position on cognition, Clark's take on PPC exemplifies how one can be a radical revisionist without being a true revolutionary. For if he is right, PPC only requires making adjustments to and inverting aspects of the traditional cognitivist picture of cognition. The radical changes he highlights fall far short of constituting a wholesale replacement of the previous framework.

4. Interestingly, despite painting PPC as "a picture of the brain as a secluded inference-machine" that relies wholly on self-evidencing, Hohwy (2014, 19) also notes the parallels with enactivist thinking, observing that "the notion of self-evidencing appears to be the epistemic cousin to the dynamic systems theory notions of self-organization and self-enabling, which are often used to explain enactivism."

5. Other theorists are even more forthright on the link between PPC and cognitivism—they regard PPC as absolutely and unavoidably wedded to the idea that the brain trades in contentful representations. Hohwy (2014) tells us, for example, that the brain's predictions about likely sensory input "*necessarily rely* on internal representations of hidden causes in the world (including the body itself)" (p. 17, emphasis added).

6. Clark (2016) is quick to observe that this sort of matchmaking is often quite piecemeal and partial, as in cases of rapid perception during which the brain only gets the "gist" of a scene (p. 27).

7. Subjects viewed one of eight possible stimulus orientations while activity was monitored in early visual areas (V1–V4 and MT+) using standard fMRI procedures. For each 16-second "trial" or stimulus block, a square-wave annular grating was presented at the specified orientation (0, 22.5 ... 157.5°), which flashed on and off every 250 milliseconds

with a randomized spatial phase to ensure that there is no mutual information between orientation and local pixel intensity.

8. This is Clark's (2016, 19) own phrase: he tells us that "the prediction task is ... a kind of bootstrap heaven." Elsewhere he says that "the impasse was solved in principle at least by the development of learning routines that made iterated visits to bootstrap heaven" (p. 20). Hohwy (2013, 15) makes a similar move, holding that "a solution to the problem of perception ... must have a bootstrapping effect such that perceptual inference and prior belief is explained as being normative in one fell swoop, without helping ourselves to the answer by going beyond the perspective of the skull-bound brain."

9. Noting this, Rosenberg (2014b, 26) also claims that "the real challenge for neuroscience is to explain how the brain stores information when it can't do so in ... sentences made up in a language of thought." This is almost right, but REC's proposed modification, as per the lessons of chapter 2, is to say that "the real challenge for neuroscience is to [get beyond trying to] explain how the brain stores information when it can't do so in ... sentences made up in a language of thought."

10. Hohwy (2013) rejects the idea that the historical constraints on how a system copes with uncertainty could be understood in terms of what he calls "mere biases." This is because he insists on construing PPC in terms of inference understood as a normative notion, where the norms in question are those of Bayesian epistemology that tells us "what we should infer given our evidence" (p. 15).

Chapter 4

1. Yet other approaches to mind and cognition—such as Cognitive Integration Theory (Menary 2007, 2015a) and Material Engagement Theory (Malafouris 2013)—are also clearly compatible with the REC framework.

2. Pushmi-pullyu representations, though primitive, possess not one but two kinds of content—both descriptive and directive contents (see Millikan 2005, 173–175).

3. We were quite explicit and clear on this point: "even the most radical of enactivists need not, and should not, deny the existence and importance of contentful and representationally based modes of thinking; it is just that these should be regarded as emerging late in phylogeny and ontogeny, being dependent on immersion in special sorts of shared practices" (Hutto and Myin 2013, 12). Thus, we also wrote: "REC happily concedes that some very important forms of cognition essentially depend on the interactions between propositional attitudes. Thus, REC is compatible with a restricted CIC. To assume otherwise is to misunderstand REC's scope and interests" (Hutto and Myin 2013, 14).

4. It is true that in both Hutto 2008 and Hutto and Myin 2013, we speak in terms of content-based forms of cognition instead of content-involving forms of cognition in places. Admittedly, with hindsight, use of that language was a mistake. Unfortunately it appears to have given rise to the idea that Hutto and Myin (2013) "subscribe to a genuinely representationalist view of language despite their radicalism about cognition" (Harvey 2015, 107; see also van den Herik 2014). While this misreading may have been fostered by our own unfortunate wording, Harvey (2015, 107) goes much further than what is implied in the text when he assumes that Hutto and Myin (2013) "reflexively conceive of everyday language use as necessarily, self-evidently representational [and that we] not only describe language and language-dependent cognition as representational, [but we] suggest that it would be ludicrous to think otherwise." These are not views we have ever endorsed. We believe there is every reason to suppose that some thoughts and speech acts are contentful in a representational sense, but we do not assume that all language is representational. Such attributions to us should not be taken seriously; after all Harvey readily admits that his evidence for ascribing them is marvelously thin because "H&M don't say much about language. What they do say is mostly in asides and parenthetical statements" (p. 107).

5. REC shows that enactivists can get by without having to fall back on a notion of informational content. Thus, *pace* Heras-Escribano, Noble, and de Pinedo (2015, 2), it does not seek to develop "new ways of understanding informational content."

Chapter 5

1. See Roy 2015 for a detailed analysis.

2. Brentano may have understated the case on this score; we agree with Roy (2015, 95) that it would be better to describe this passage as a "remarkably ambiguous text."

3. See Menary 2009 for a recent exegesis of Brentano's ideas.

4. Famously, Searle insists, "Language is derived from Intentionality and not conversely" (p. 5; see also Searle 2011).

5. As Roy (2015, 94) also points out, a clear sign that we lack an illuminating analysis of intentionality is that the concept is "frequently paraphrased with the help of a set of recurring notions such as contentfulness (an intentional state is a state with a content), directedness (an intentional state is a state directed at something), reference (an intentional state is a state referring to something), attitude (an intentional state is an attitudinal state), or satisfiability (an intentional state is a state with conditions of satisfaction)."

6. No doubt, old habits die hard. Still, Muller (2014) is surely right in his diagnosis of what has been holding back our understanding of alternate ways of thinking about intentionality. He writes: "The persistent and consistent focus on the propositional attitudes in general, and belief and desire in particular, has habituated us into a narrow way of thinking about representation within the philosophy of mind" (p. 176).

7. We only report Crane's (2009) view here; we do not suggest that it is unproblematic. An immediate worry about the proposal is that it is hard to understand what it would be for an experience or picture to be accurate simpliciter. It looks as if to be accurate is to be accurate in this or that respect. And that would seem to imply that a condition on being an experience or picture is that the ways in which the picture or experience might be accurate or not would have to be independently specifiable. If so, that makes it seem that for an experience or picture to be accurate or not (to whatever degree) depends on its being used for a particular purpose that determines in what respect it might be so.

8. Back in the day, Hutto (1999) endorsed a vision of intentionality along precisely these lines holding that intentionality entails the existence of correctness conditions that differ from truth conditions.

9. Making room for nonrepresentationalism in the current intellectual climate is going to be a long, hard battle in some quarters. Rey (2002, 405), for example, confronts Dreyfus with the fact that he finds it "hard to understand how a brain can respond to 'solicitation' without representing it as a solicitation." Of course, the only way to reply to those who flaunt their imaginative difficulties as if they constitute a sound basis for arguments is to ask what legitimizes the pictures that impose such limits on the imagination. Once one neutralizes the authority of such restrictive pictures and assumptions, it really isn't very difficult to imagine that an organism might respond to some particular kind of solicitation, even in quite sophisticated ways, without representing it as a solicitation.

10. In this vein, Dreyfus (2002b, 415) insists that the "holistic response is no reflex but there is no need in this account for the sort of explicit mental representations involved in planning."

11. For a discussion of the sort of confusion that can ensue see Hutto 2014. It should be noted that in making this terminological recommendation, we have no desire to try to police philosophical language. After all, as Jackson and Pettit (1993, 269) point out, "'Content' is a recently prominent term of art and may well mean different things to different practitioners of the art." Our choice of terminology is simply driven by the desire to be maximally clear about exactly which properties are being assumed to characterize the most basic forms of intentionality.

12. Other nonrepresentationalist proposals are not deemed to escape the fundamental dilemma. For example, Roy (2015b, 118–119) asks, "Does Rowlands' alternative conception of the nature of intentionality manage to break away from such a representationalist characterization? … The answer is negative and without much possibility for appeal, for his alternative conception turns out to be even more committed to the idea of objective specification than the neo-Brentanian one." As Roy (2015b) highlights, "The insistence with which Mark Rowlands

emphasizes that a representation is primarily something that 'makes claims about the world' is, for instance, clearly consonant with ... the thesis that representation is essentially specification" (p. 110). See Rowlands 2015b for more evidence of the strength of his commitment to the representationalist specification requirement.

13. For this reason, the mechanisms that distribute pigmentation so as to match the skin of chameleons to their immediate environments are not candidates for being intentionally directed. This is because such devices fail to satisfy the final condition.

14. Indeed, as noted in chapter 2, some—such as Miłkowski (2015)—still hold out hope that it has already provided what is needed for dealing with the HPC.

15. Millikan (1993) is beautifully clear on this important point. She writes that "having a certain history is not, of course, an attribute that has 'causal powers' ... that a thing has a teleofunction is a causally impotent fact about it" (Millikan 1993, 186).

16. Shea (2013, 498) reports that on the standard view, "Information processing theories effectively offer a wiring diagram showing how inputs affect states of the system and, in conjunction with other states of the system, issue in behavioural outputs. What does it add to that wiring diagram to label various nodes with representational contents? A realist about mental representation is committed to the reality of the internal particulars described in the theory, and of their contents."

17. The same verdict applies equally to linguistic contents. Thus Millikan (2005, 87) asks: "What objective criterion determines that one is using a dog thought only in response to a dog or that one's dog thoughts always correspond even to the same kind of thing? I adopt Sellars's suggestion that adequate intentional representing is a kind of picturing or mapping."

18. Millikan expressed the same view to Hutto privately in the following terms "Connecting with something black-and-a-dot is no part of any proximate normal explanation of why it helped its ancestors survive. Neither the blackness nor the dotness helped in any way; neither

need be mentioned. But the nutritious object was essential" (Millikan, personal communication, 1996).

19. Associating REC's view of Ur-intentionality with Targeted Response Tendencies, Kiverstein and Rietveld (2015, 709) claim that REC's take on basic cognition "implies a problematic view of animal behaviour as either hard-wired or learned dispositions to respond to fixed and stable environmental cues. The selective responsiveness of animals ... is thereby conceived of as the animal being equipped with mechanisms that are set off only by specific environmental triggers."

20. Where we disagree with Kiverstein and Rietveld (2015, 712) is in their claim that "skill is exercised in acting so as to improve one's grip on a situation and one's sense of what the situation demands from one as one acts." As per our previous concerns about Noë's (2004) sensorimotor account of enactivism, see Hutto and Myin 2013, chap. 2. We are suspicious of the claim that a "practical understanding" is required to mediate and gives access to possible actions.

Chapter 6

1. Accepting that some forms of thought and language involve reference and truth conditions is not, of course, to endorse the reductive thesis that all uses of language must be contentful in this sense.

2. In an extreme statement, meant to capture the options available in this apparently forced choice, Korbak (2015, 95) maintains that either "basic cognitive systems do genuinely communicate because they are alive, or nothing (not even H&M themselves) does."

3. Philosophical criticisms of extreme naturalism are easy to find. For example, Williamson (2014a, 37) constructs the following quick but effective argument against extreme naturalism: "If it is true that all truths are discoverable by hard science then it is discoverable by hard science that all truths are discoverable by hard science. But it is not discoverable by hard science that all truths are discoverable by hard science. Therefore the extreme naturalist claim is not true. ... Truth is a logical or semantic property, discoverability an epistemic one, and a hard science a social process."

4. This comment is made as an aside, in a note, but it raises an important challenge for REC if true. Accusations of discontinuity are, in any case, a running theme in Menary's work. Menary (2009) identifies Brentano's formulation of intentionality as a main source of the unwelcome idea that minds are discontinuous with the rest of the world.

5. Teleosemantic theories of content aim to show how "teleology turns into truth conditions" (McGinn 1989, 148). For example, in Millikan's version, very roughly, a device has the teleofunction of representing Xs if it is used, interpreted or consumed by the system because it has the proper function of representing the presence of Xs. Proper functions are called on to explain how it is that content is fixed by what organisms are supposed to do in their consumptive activity, as opposed to what they are merely disposed to do. Sterelny (2015, 552) reminds us that "in general, teleosemanticists are representational liberals: they are happy to attribute content to the control systems of the simplest of organisms—to bacteria with their magnetosomes—and to the simple subsystems of more complex agents. For even the simplest organisms systematically register and respond adaptively to some features of their environment: they have control states that vary by design with states of their normal environment and direct behaviour that is appropriate, or would be appropriate, were the environment to be in the state to which the control state is supposed to be tuned (for the correlation is imperfect of course)."

6. Hence, cognitive activity "can be understood as falling along a continuum: far down at the non-semantic end, we have interactions that are best explained in terms of their physical properties. On the semantic end, we have interactions that are best explained in terms of their semantic properties" (Cao 2012, 55).

7. The Scaling Down Objection, as Korbak (2015, 92–93) characterizes it, is an inversion of the scope objection—the criticism that explanations given in terms of contentless cognition won't scale up (for a discussion see Hutto and Myin 2013, 45ff.).

8. For a detailed discussion of how the REC take on the transformative powers of scaffolded minds compares to and differs from that of McDowell, see Hutto 2006a.

9. Here we focus exclusively on concerns that REC promotes what Bar-On (2013) identifies as a radical, deep-chasm, diachronic version of continuity skepticism. This is because the synchronic variant of such skepticism, which focuses on the gaps between currently existing cognitive creatures, is entirely compatible with the presence of a full range of intermediaries, with no missing links, over the course of natural history.

10. For more on how to understand the capacities for learning and teaching as species-relative biologically inherited capacities, see Tomasello 1999; Rakozcy, Warneken, and Tomasello 2008, 2009; Csibra 2009.

11. O'Brien and Opie (2015) are indelibly clear in making this charge and in promoting a CIC alternative: "There is a fundamental problem with the idea of contentless intentionality: it's been tried before, and it doesn't work. Back then the scheme was known as 'behaviourism,' rather than 'targeted directedness,' but the two ideas are of a piece. Behaviourists sought to explain animal behaviour, including all the complexities of human problem solving and language, in terms of the history of stimulus-response events to which organisms (of each kind) are typically exposed. The bankruptcy of this approach consists in the fact that moment-by-moment stimuli are simply too impoverished to account for the richness, variety, and specificity of the behaviours that animals exhibit. *It just isn't possible* to explain the ability of evolved creatures to selectively engage with features of the environment—in other words, engage in targeted behaviour—*without supposing they employ internal states that in some way represent those features.* ... [The NOC program, and by implication REC,] misunderstands the broader explanatory project [which is to explain] ... intelligence rather than just intentionality" (p. 724, emphasis added). Kiverstein and Rietveld (2015) are slightly more cagey in accusing REC of endorsing nothing more than stimulus-response behaviorism, but they too think its vision of Ur-intentionality "implies a problematic view of animal behaviour as either hard-wired or learned dispositions to respond to fixed and stable environmental cues. The selective responsiveness of animal behaviour which we have taken to be constitutive of skills is thereby conceived of

as the animal being equipped with mechanisms that are set off only by specific environmental triggers" (p. 708).

12. Mindreading is defined here as a process that involves the attribution of mental contents in ways that entail the manipulation of contentful mental representations—whether theoretical inferences, simulations, or both.

13. In defending REC on precisely this score, Medina (2013, 324) relates that for Bar-On (2013), "Expressive behavior is far richer than mere 'brute signaling' or causally produced indication."

14. These short replies and observations only scratch the surface. For a fuller argument that discusses empirical findings about infant cognition, see Hutto 2015a.

15. Tomasello (1999, 4) has long defended the idea that "the amazing suite of cognitive skills and products displayed by modern humans is the result of some sort of species-unique mode or modes of cultural transmission." But see Satne 2014 for a REC-friendly critique of Tomasello's CICish cognitivism.

Chapter 7

1. The broad use we focus on is the one found in countless cognitive science textbooks. Rey (2015, 171) provides a familiar statement: "'Representation' has come to be used in contemporary philosophy and cognitive science as an umbrella term to include not only pictures and maps, but words, clauses, sentences, ideas, concepts, indeed, virtually anything that is a vehicle for intentionality (i.e. anything that stands for, 'means', 'refers to', or 'is about something')." Rey commits to the broad usage we target in saying that "something represents whatever it represents—namely has the representational content that it has."

2. For a general discussion of hyperintellectualism and arguments against it see Hutto 2005; Hutto and Myin 2013, chap. 5; Myin and Degenaar 2014.

3. Rescorla (2016), for example, is very clear on this point: "Bayesian sensorimotor modelling requires that the motor system conform (at least approximately) to Bayesian norms. It does not require that the motor system represent Bayesian norms. For example, it does not require that the motor system represent Bayes's Rule" (p. 12). Similarly, Orlandi (2014, 3) says: "A system may have features—for example, wires or constraints—that developed, and continue to develop, under evolutionary and environmental pressure. These features have a certain function. They make the system act lawfully; that is, they make the system act in a way that is describable by principles. The principles, however, are in no sense represented by the system and encoded within it."

4. Rescorla (2016) uses the "science says/does" illocution no less than fifteen times in a single article, reporting what the science: describes (pp. 4, 27, 28); posits (p. 17); remains neutral about (p. 20); explains (p. 22); illuminates (p. 22); cites (p. 23); assigns (p. 23); may not vindicate (p. 25); does not employ (p. 27); does not advance (p. 28); aims to provide (p. 29); purports (p. 29); and extends (p. 31).

5. It is, of course, possible to insist that this circularity is entirely virtuous, as Burge (2010) tries to do. In Hutto and Myin 2013, chapter 6, we argue that this sort of philosophical appeal to what science says or ought to say, especially when the science in question is unfinished and still evolving, is not only a poor way to try to justify philosophical claims about the mental representation, if taken seriously it could actually retard scientific progress.

6. In making this proposal Gładziejewski (2015, 2016) hopes to address Ramsey's (2007) job-description challenge, and deal with the worry that representational constructs used in the cognitive sciences are often representational in name alone—that is, that representational terminology too often serves as only "an empty and misleading ornament." As he puts it, "It is easy to say that representations are component parts of mechanisms that play the functional role of a representation. But it is much harder to answer the question of what it means to function as a representation within a mechanism. When are we justified in attributing the role of a representation to a component of a neural or computational mechanism? What exactly does a component have to do within a

mechanism in order to be justifiably categorized as a representation?" (Gładziejewski 2015 67).

7. Gladziejewski (2015) identifies Grush's (1997, 2004) emulator theory of representations—which holds that motor control perception and imagery make use of internal emulators of the body and world—as a shining example of the sort of theory that posits constructs that comfortably satisfy the four criteria that must be met, in his view, for items to count as representations (Gladziejewski 2015, 85).

8. It is useful to think about how keys and locks work in this regard. Assume that every mental structure has its own unique structural properties or geometry. Accordingly, the "shape" of such structures and how they interact in the machinery of the mind are what drive cognition. It is easy to see how the analog properties of a given structure plausibly "determine the causes and effects of its tokenings in much the way the geometry of a key determines which locks it will open" (Fodor 1991, 41). It is not obvious how any putative content such a structure might bear could do likewise, or indeed why the existence of items with these sorts of properties should be thought of as intrinsically contentful.

9. Although Orlandi (2014) focuses solely on vision, we are assuming that lessons we draw here apply to all of perceptual science.

10. Mole and Zhao (2015) also spot this inconsistency in Orlandi's (2014) thinking, and they argue for representationalism about both processes and products.

11. Motivating this proposal, Gerrans (2014, 14) stresses that "collecting and collating correlations between neural, phenomenological and cognitive properties ... is useful but we need a theoretical approach that *fits all this information together.*"

12. Accordingly, persons are conceived of "as complex, hierarchically-organized information-processing systems implemented in neural wetware" (Gerrans 2014, 16).

13. The autonomy thesis endorses the irreducibility claim about the mind of the sort Davidson (1987, 46) championed long ago: "The reason mental concepts cannot be reduced to physical concepts is the

normative character of mental concepts." The autonomy thesis about the mental can be understood in more or less realistic terms. Yet in all versions the root idea is that propositional attitudes can only be ascribed, or only have life, when they stand in appropriate rational relations.

14. For example, Gerrans (2014) insists that proponents of the autonomy thesis must embrace the idea that organic damage might "play a causal role in introducing … drastic change in psychological structure but plays no explanatory role" (Gerrans 2014, 27, emphasis added).

15. This integrationist vision of the cognitive is clearly in tune with the idea that "The whole thrust of cognitive science is that there are sub-personal contents and sub-personal operations that are truly cognitive in the sense that these operations can be properly *explained only in terms of these contents*" (Seager 2000, 27, emphasis added).

16. For example, he elsewhere speaks of the "*necessary* role of cognitive theory in linking the neurobiological and phenomenological levels of explanation" (Gerrans 2014, 18, emphasis added). Methodologically speaking, it is strange that Gerrans (2014) appeals to such general and wholesale philosophical justifications, given that he pins his philosophical colors so strongly to the naturalist mast.

17. Enactivists, of course, encourage multistranded investigations, involving explanations that are pitched at various "levels" and "scales." Gerrans (2014) acknowledges this. Taking the case of vision as a prime example, he emphasizes the need for theories that seek to simultaneously investigate different levels of cognitive activity and how they integrate. Yet here he notes that "even enactive theorists of vision who disagree with Marrians nonetheless debate with them about the causal relevance of mechanisms at different levels" (p. 43).

Chapter 8

1. Clowes and Mendonça (2016) propose that fundamentally non-representational kinds of cognitive equipment might be reused for representational purposes in such representation-hungry contexts. See Degenaar and Myin 2014 for an extended, REC-friendly argument

against the need to think of such contexts as representation hungry in the first place.

2. Machery (2009, 222), for example, dismisses the tenability of non-representational accounts of even the most basic kinds of cognition on just these grounds: "Grush has argued that physical actions are often guided by representation of feedback ... so even simple actions cannot be explained without positing representations."

3. Gerrans (2014) also says that imagination deploys specialized neural circuitry to "construct and manipulate *representations which have representational contents but no congruence conditions*" (p. 114, emphasis added). Anyone who holds that mental content requires and is indeed defined by its correctness or congruence condition will find it utterly puzzling that Gerrans (2014) also claims that imaginings have representational contents despite the fact that they lack congruence and correctness conditions altogether. What remains of representational content if you subtract congruence conditions from a mental representation? Gerrans's (2014) answer is intentional structure. But it is not clear what, for cognitivists, could put the intentional into such structure if not the existence of mental representations with congruence conditions.

4. Speaking on behalf of REC, Medina (2013, 318) makes clear that "the enactivist account ... does not deny that our imaginings often involve representational capacities and representational contents—indeed, this is clearly the case in sophisticated exercises of the imagination such as watching a movie or reading a novel. But, even in these cases, the representational elements of the imagination cannot account for many aspects of the imaginative experience. ... There are forms that the imagination can take that do not require representational contents at all: ways of acting and interacting imaginatively without representing what one is imaginatively enacting or re-enacting."

5. In fact, the as-possible view of the correctness conditions of imaginings, as presented in Langland-Hassan 2015, is a variation on Yablo 1993, which holds that "to find p conceivable is to be in a state which (i) is veridical only if possibly p, and (ii) moves you to believe that p is possible" (p. 7).

6. On Langland-Hassan's (2015) account, hybrid imaginative attitudes qualify as sensory imaginings if they have sensory images as proper parts.

7. One reason for thinking that both components should be contentful is that this would help us to better understand how they could combine. If both imagistic and discursive components are contentful, this might be thought to help with the problem of clashing formats: "It is generally assumed that these two formats of representation are like oil and water—they don't mix" (Langland-Hassan 2015, 17). But in fact, given that the devil will be in the details and given that we lack a detailed naturalistic theory of content, it is hard to see how the mere fact that both elements are deemed contentful addresses the clashing-formats problem. It remains a challenge for any theory that grants a role for imagistic thought in practical reasoning to "confront the issue of how imagistic thought—if it does indeed occur in a non-discursive, iconic format—inferentially interacts with discursive thought" (Langland-Hassan 2015, 18).

8. Langland-Hassan's (2015) hybrid account is said to allow us "to arrive at a workable, naturalistic account of imagination's correctness conditions" (p. 17). It is not clear how this is so given that the field currently lacks any convincing naturalistic theory of content, and especially in light of the fact that Langland-Hassan himself does "not pretend to offer any answers" (p. 17).

9. The term "hominin" denotes any species of early human that is more closely related to humans than chimpanzees, including modern humans.

10. For this very reason, hominin toolmaking has been held up as the clearest evidence of nonverbal mediated instrumental reasoning (Bermúdez 2003).

11. Embodied intentions of the kind that toolmaking involves are instances of extensive, and not merely extended, forms of cognition. For an articulation of this distinction see Hutto, Kirchhoff, and Myin 2014.

12. As described, Foglia and Grush (2011) depict a hybrid task—requiring adopting imaginative attitudes that are in part contentful judgments. The subjects must manipulate the shape to judge if the shapes are congruent. But we can imagine a basic version of the task that even hominins might engage in. It might be that fitting a particular flake into another stone might demand determining solely by basic, noncontentful imaginative means if the stone and the destined place have congruent shapes without being able to describe the task in such terms at any level. Presumably one might engage in such an exercise, driven by the task requirements, even if one lacked the concept of congruence.

13. It is important to note that even if it could be shown that basic imaginings have representational content, a further argument would be required to show that such content makes an explanatorily contribution—that it isn't a metaphysical dangler.

Chapter 9

1. In 1972 Tulving speculated there were likely around fifty or so categories of memory at large in psychology. Famously, he added two of the best-known categories, episodic and semantic memory, to that list (see Roediger 1980, 235). That was then: today Michaelian (2016, 18) speaks of there being "hundreds of *types* of memory distinguished by psychologists."

2. There are compelling empirical reasons for doubting that any singular memory system gives rise to the various forms of remembering. In this light it has been doubted that "memory" constitutes a natural kind. After a careful review of the evidence, Michaelian (2016) concludes that "while declarative memory may be a natural kind, memory as a whole—including both declarative and nondeclarative memory—is not. ... [Hence] it is unlikely to be possible to develop a unified account of remembering and memory knowledge in general" (p. 17; see also Michaelian 2011).

3. Such remembering can be voluntarily prompted or it might simply be triggered by sensory means—by encountering particular looks, smells, tastes, or feels linked with familiar people, places, or things.

4. Although we can potentially narrate any specific event or recurring events in our lives, including acts of remembering themselves, only one special sort of memory—autobiographical reverie—has a strong claim for being indelibly narrative in nature.

5. SIT has considerable empirical backing from a wealth of findings "accumulated in psychological experiments over more than 100 years" (Nelson 2007, 185). At a minimum it is supported by "research on parent-child reminiscing that extends over more than a 20-year period" (Fivush and Nelson 2006, 235). Yet as Sutton (2007, 82) observes, despite this interesting work in psychology in trying to understand the links between narrative practices and memory, "Only a few philosophers ... have looked to the psychology of autobiographical memory for understanding of the constraints on our contact with the personal past." Given SIT's strong claims and robust empirical backing, it is perhaps surprising that it is not more widely discussed in philosophy. A plausible explanation of this fact appears to be that it is another casualty of the general tendency of philosophers—especially those in some wings of the analytic tradition—to assume that the essence of phenomena can be investigated independently of science (for further discussion of this tendency, see the preface to Michaelian 2016).

6. Or as Fivush and Nelson (2006, 236) put it, "The developmental 'mechanism' requires linguistically mediated social interaction."

7. Other support for the idea that the narratives in question must be linguistic stems from the plausible hypothesis that a command of verbal narratives, whether oral or written, is necessary for representing reasons, motives, and actions and thus for providing the "means for organizing sectors of experience into intentional systems ... that have the profile of persons" (Herman 2013, 23). It is narratives of this unique sort, those with this particular subject matter—aka person narratives or folk psychological narratives—that have "special features and a special role in our individual and collective lives" (Currie 2010, 219; see also Goldie 2004 and Hutto 2008).

8. In defending this idea Nelson (2003, 28) identifies "three essential components of narrative [that] ... remain weakly present or non-

existent in most of the narrative productions of the 3- to 5-year-old preschooler: temporal perspective, the mental as well as physical perspective of self and of different others, and essential cultural knowledge of the unexperienced world."

9. Nelson (2003) discusses this phenomenon with reference to work by Miller et al. (1990) that provides "observations of appropriation by a child from a parent's account. For example, one small boy who was told by his mother in a cautionary way about having fallen off a stool when she was young retold this story later as his 'when he fell off the stool.' Miller also reports children in preschool freely appropriating" (p. 31).

10. One barrier to an early onset of this capacity is that "at 2 years of age the child has only a few of the rudiments that enter into narrative" (Nelson 2003, 27). At the early stages in the child's life, "language is being learned and used but it is not yet a vehicle for conveying the representation of narrative" (p. 27).

11. The claim that the mature development of narrative skills makes it possible for people to remember events in their lives in autobiographical ways should be distinguished from claims that such narrative capacities are responsible for generating a narrative sense of self and the even stronger claim that they are a basis for self-constitution. Schechtman (1996, 2007) defends the narrative self-constitution view, but see Strawson 2004 and Hutto 2016 for skeptical arguments.

12. Developmental psychologists reliably report that "the period from 3 to 5 years appears to be one especially crucial phase of transition in the development of children's conceptions of persons" (Richner and Nicolopoulou 2001, 398).

13. That memory is fundamentally creative and reconstructive is not a new idea. As Schacter et al. (2012, 681) remind us, "The general idea that memory is a constructive process ... rather than a literal replay of the past, dates to the pioneering work of Bartlett (1932), and has been developed by a variety of investigators who have demonstrated the occurrence of memory distortions and theorized about their basis."

14. In 2007 the neuroscientific discoveries that revealed strong links between memory and imagination made the journal *Science*'s "list of the top ten discoveries of the year" (*Science*, December 21, 2007, 1848–1849, as reported in Schacter et al. 2012, 677).

15. This conclusion, however, must be tempered by other findings that complicate the story to a significant extent. Despite the repeated empirical demonstrations of "impressive similarities between remembering the past and imagining the future, theoretically important differences have also emerged" (Schacter et al. 2012, 678). Any fully satisfactory account of the cognitive basis of mental time travel will need to take stock of and accommodate these findings about documented differences between the relevant acts of remembering and imagining. This important constraint on any future theorizing highlights the urgent need for the development of a sufficiently nuanced integrative theory of the cognitive basis of mental time travel (see Schacter et al. 2012, 683).

16. Citing research by Pezdek, Blandon-Gitlin, and Gabbay (2006), De Brigard (2014, 163) reports that "most ordinary cases of misremembering have an air of plausibility to them."

17. As such, acquiring a sense of self as a person with a particular past is "dependent upon language used to exchange views about self and other, primarily through narratives but also through commentary on the self by others, as well as on their own feelings, thoughts and expectations of what might happen" (Nelson 2003, 33). Thus acquiring a sense of self of the sort needed for autobiographical remembering is something that happens "during the critical years when the child can enter fully into the linguistic world but is not yet a participant in formal schooling" (Nelson 2003, 22).

18. Various philosophers have defended the thesis that contentful thought and talk depend on the mastery of special linguistic practices on purely conceptual grounds (see, e.g., Davidson 1984; Price 2013). For a naturalistically motivated defense of this idea see Hutto and Satne 2015.

19. For example, Gerrans (2014) maintains that "*qua* simulations imaginative states do not have congruence conditions" (p. 105; see also p.

18). Langland-Hassan (2015, 665) tells us that "much of what has been said about sensory imagination conflicts with the idea that imaginings have substantive correctness (or veridicality, or accuracy) conditions at all." Although Michaelian (2016) holds that reconstructive memory retrieval trades in representations with sensory content, he allows that such content is "nonpropositional, in the sense that it cannot be evaluated for truth and falsity in a binary manner" (p. 53). Nevertheless, he notes that it can have propositional elements and does have accuracy conditions. For a more detailed discussion of the question of imaginative content, see Medina 2013 and Hutto 2015a.

20. The content-based approach has an ancient pedigree that can be traced back to Aristotle in the Western tradition at least. Undoubtedly, it—aka memory representationalism—is the "default philosophical view" (De Brigard 2014, 159). Campbell (2006) dubs this view the "archival picture of memory," according to which "memory is depicted as the capacity to make detailed mental representations of our experiences, and then to store these representations discretely and in some manner that allows us to call them to mind on subsequent occasions" (p. 362).

21. Citing work by Robinson and Swanson (1993) and McIsaac and Eich (2002), Sutton (2010, 29) speaks of a robust consensus "developing about certain systematic properties of field and observer perspectives in memory. Some consistent findings are that field perspectives are in general significantly more common, but that observer perspectives—always found in a significant minority of memories—increase in memories of more temporally remote events; that memories recalled from an observer perspective tend to include less affective and sensory detail than those recalled from a field perspective; and that memories are more likely to be recalled from an observer perspective when the person either was more self-conscious or self-aware during the original experience, or is more self-conscious or self-aware at the time of recollection."

22. The chances of error increase over time. Thus De Brigard (2014, 161) reports that "Loftus et al. (1978) discovered that when the interview was administered 20 min after witnessing the event, participants

were correct about 40% of the time, but if the interview is administered one week after witnessing the event, the rate dropped to 18%."

23. Subjects are "more confident saying that the events they did not imagine definitely did not happen in their lifetime than they were about the events they just imagined" (De Brigard 2014, 161).

24. There is also evidence, which ought to be surprising for the content-based view, that "individuals with memory-related pathologies tend to misremember less than normal subjects" (De Brigard 2014, 162).

25. Motivated by these findings, De Brigard (2014, 167) claims that "a better alternative ... is to interpret the frequency of memory distortions ... as the ... byproduct of a mechanism that is actually doing something else." In particular, he maintains that we have every reason to believe that "many ordinary cases of misremembering ... [are] ... the normal result of a larger cognitive system that performs a different function" (p. 158).

26. As Goldie (2012, 152–153) reminds us, "Fictional narratives do not aspire to be true, whereas real life narratives do. A narrative is fictional not in virtue of its content being false, but in virtue of its being narrated, and read or heard, *as part of a practice of a special sort.*" Highlighting this distinction, Goldie stresses that "reference and truth have no application in fiction, but do have application in historical and everyday explanation" (p. 154).

27. We must guard against leaping to the conclusion that constructive memory processes necessarily lead to distortion and error. For as Michaelian (2016) warns, "It might seem obvious that such incorporation can only decrease the accuracy of memory. But the typical experimental setup in this area focuses precisely on providing subjects with misleading post-event information. If it turns out that, in ecological settings, subjects are more likely to receive accurate than inaccurate post-event information, then the incorporation of testimonial information may actually increase the accuracy of retrieved memories, despite resulting in retrieved memories that depart from the experience of events" (pp. 10–11; see also Sutton 2010).

28. Campbell (2006, 377) is surely right that recollection is a complex social activity, one in which "taking care to get the past right involves at least our implicit accountability to others." In this sense, determining "memory's faithfulness to the past is, in many cases, a complex epistemological/ethical achievement" (p. 363).

Chapter 10

1. Michaelian (2016) is not the only one of whom, on this pivotal issue, RECers might ask "Et tu, Brute?" Even Nelson (2007) takes such a view. This is seen when she draws a contrast between autobiographical and other kinds of memory. She says the latter "evolved to preserve information learned in encounters with the environment; they are useful in predicting new encounters and taking effective action within them (Nelson 1993, 2003). But these memorial contents do not include consciousness of self in the past encounters" (Nelson 2007, 187).

2. Thus REC resists the idea that cognition "should be understood as the construction of information as part of a 'user-interface' that enables us to interact with the world … [where] this could suggest a profoundly different framework for thinking about information" (de-Wit et al. 2016, 10).

3. Wheeler (2015, 13) adds, "But if this is right, then not only are considerations of relationality poorly placed to support the extensive mind over the extended mind, they are equally poorly placed to support externalism regarding our cognitive machinery over internalism."

4. As discussed in chapter 2, if there is no content then there are no vehicles of content. The two notions are conceptually linked. Of course, one might attempt to preserve the label *vehicle*, by using the term to designate, say, as Wheeler (2015, 2) does, the "material realizers of cognition." But that would no longer be a vehicle in the traditional sense and one would need some alternative, nonsemantic criterion for demarcating what counts as such a material realizer.

5. As Adams and Aizawa (2010, 72) acknowledge, "Given the state of current science, we only identify a person's brain states [as cognitive] via inferences to the content of those states."

6. Craver (2007, 141) also mentions that some cognitive mechanisms make use of resources outside organismic boundaries to such an extent that "it may not be fruitful to see the skin, or the surface of the CNS, as a useful boundary."

7. Wheeler (2015, 3) acknowledges that "extended functionalism needs to be augmented by a mark of the cognitive, in order to deliver more than the conceptual possibility of cognitive extension." He tells us that he is "'appropriately poised' to deliver it." We await details.

References

Abramova, K., and M. Villalobos. 2015. The apparent (ur)intentionality of living beings and the game of content. *Philosophia* 43:651–668.

Adams, F., and K. Aizawa. 2010. Defending the bounds of cognition. In *The Extended Mind*, ed. R. Menary. MIT Press.

Aizawa, K. 2014. The enactivist revolution. *Avant (Torun)* 5 (2): 19–42.

Aizawa, K. 2015. What is this cognition that is supposed to be embodied? *Philosophical Psychology* 28 (6): 755–775.

Akins, K. 1996. Of sensory systems and the "aboutness" of mental states. *Journal of Philosophy* 93 (7): 337–372.

Alksnis, N. 2015. A dilemma or a challenge? Assessing the all-star team in a wider context. *Philosophia* 43:669–685.

Alsmith, A. J. T., and F. de Vignemont. 2012. Embodying the mind and representing the body. *Review of Philosophy and Psychology* 3 (1): 1–13.

Anderson, M. L. 2010. Neural reuse: A fundamental organizational principle of the brain. *Behavioral and Brain Sciences* 33:245–266.

Anderson, M. L. 2014. *After Phrenology: Neural Reuse and the Interactive Brain*. MIT Press.

Andres, M., X. Seron, and E. Olivier. 2007. Contribution of hand motor circuits counting. *Journal of Cognitive Neuroscience* 19:563–576.

Barandiaran, X., and E. Di Paolo. 2014. A genealogical map of the concept of habit. *Frontiers in Human Neuroscience* 8:522.

Barnier, A. J., and J. Sutton. 2008. From individual to collective memory: Theoretical and empirical perspectives. *Memory* 16 (3): 177–182.

Bar-On, D. 2013. Expressive communication and continuity skepticism. *Journal of Philosophy* 110 (6): 293–330.

Barth, C. 2011. *Objectivity and the Language-Dependence of Thought: A Transcendental Defence of Universal Lingualism*. Routledge.

Bartlett, F. C. 1932. *Remembering: A Study in Experimental and Social Psychology*. Cambridge University Press.

Bechtel, W. 2008. *Mental Mechanisms: Philosophical Perspectives on Cognitive Neuroscience*. Routledge.

Bechtel, W. 2016. Investigating neural representations: The tale of place cells. *Synthese* 193:1287–1321.

Bechtel, W., and R. C. Richardson. [1993] 2010. *Discovering Complexity: Decomposition and Localization as Strategies in Scientific Research*. 2nd ed. MIT Press.

Beer, R. D. 2000. Dynamical approaches to cognitive science. *Trends in Cognitive Sciences* 4 (3): 91–99.

Beer, R. D. 2003. The dynamics of active categorical perception in an evolved model agent. *Adaptive Behavior* 11:209–243.

Bermúdez, J. 2003. *Thinking without Words*. Oxford University Press.

Brainerd, C. J., and V. F. Reyna. 2005. *The Science of False Memory*. Oxford University Press.

Branquinho, J. 2001. *The Foundations of Cognitive Science*. Oxford University Press.

Brentano, F. C. [1874] 2009. *Psychology from an Empirical Standpoint*. Routledge & Kegan Paul.

Brogaard, B. 2014. Does perception have content? In *Does Perception Have Content?*, ed. B. Brogaard. Oxford University Press.

Brook, A. 2007. Introduction. In *The Prehistory of Cognitive Science*, ed. A. Brook. Palgrave Macmillan.

Bruineberg, J., and E. Rietveld. 2014. Self-organization, free energy minimization, and optimal grip on a field of affordances. *Frontiers in Human Neuroscience* 8:1–14.

Burge, T. 2010. *The Origins of Objectivity*. Oxford University Press.

Byrge, L., O. Sporns, and L. B. Smith. 2014. Developmental process emerges from extended brain-body-behaviour networks. *Trends in Cognitive Sciences* 18:395–403.

Campbell, D. 2014. Review of *Radicalizing Enactivism* by D. D. Hutto and E. Myin. *Analysis* 74 (1): 174–176.

Campbell, J. 2008. Causation in psychiatry. In *Philosophical Issues in Psychiatry,* eds. K. Kendler. and J. Parnas. Johns Hopkins University Press.

Campbell, S. 2006. Our faithfulness to the past: Reconstructing memory value. *Philosophical Psychology* 19 (3): 361–380.

Cao, R. 2012. A teleosemantic approach to information in the brain. *Biology & Philosophy* 27 (1): 49–71.

Cappuccio, M., and T. Froese. 2014. Introduction to making sense of non-sense. In *Enactive Cognition at the Edge of Sense-Making*, ed. M. Cappuccio and T. Froese. Palgrave Macmillan.

Carruthers, P. 2009. Invertebrate concepts confront the generality constraint (and win). In *The Philosophy of Animal Minds*, ed. R. W. Lurz. Cambridge University Press.

Carruthers, P. 2011. *The Opacity of Mind: An Integrative Theory of Self-Knowledge*. Oxford University Press.

Casasanto, D., and K. Dijkstra. 2010. Motor action and emotional memory. *Cognition* 115 (1): 179–185.

Casey, E. 1987. *Remembering: A Phenomenological Study*. Indiana University Press.

Cash, M. 2008. Thoughts and oughts. *Philosophical Explorations* 11 (2): 93–119.

Chalmers, D. 1996. *The Conscious Mind*. Oxford University Press.

Chambers, D., and D. Reisberg. 1985. Can mental images be ambiguous? *Journal of Experimental Psychology: Human Perception and Performance* 11:317–332.

Chemero, A. 2009. *Radical Embodied Cognitive Science*. MIT Press.

Chomsky, N. 2007. Language and thought: Descartes and some reflections on venerable themes. In *The Prehistory of Cognitive Science*, ed. A. Brook. Palgrave Macmillan.

Chow, J. Y., K. Davids, C. Button, and I. Renshaw. 2015. *Nonlinear Pedagogy in Skill Acquisition: An Introduction*. Routledge.

Chow, J. Y., K. Davids, R. Hristovski, D. Araújo, and P. Passos. 2011. Nonlinear pedagogy: Learning design for self-organizing neurobiological systems. *New Ideas in Psychology* 29:189–200.

Churchland, P. M. 1979. *Scientific Realism and the Plasticity of Mind*. Cambridge University Press.

Churchland, P. M. 1993. Evaluating our self conception. *Mind & Language* 8 (2): 211–222.

Clapin, H., ed. 2002. *Philosophy of Mental Representation*. Clarendon Press.

Clark, A. 1997. *Being There: Putting Brain, Body, and World Together Again*. MIT Press.

Clark, A. 2003. *Natural-Born Cyborgs: Minds, Technologies and the Future of Human Intelligence*. Oxford University Press.

Clark, A. 2006a. Language, embodiment and the cognitive niche. *Trends in Cognitive Sciences* 10 (8): 370–374.

Clark, A. 2006b. Material symbols. *Philosophical Psychology* 19 (3): 291–307.

Clark, A. 2008a. Pressing the flesh: A tension in the study of the embodied, embedded mind? *Philosophy and Phenomenological Research* 76:37–59.

Clark, A. 2008b. *Supersizing the Mind: Embodiment, Action, and Cognitive Extension*. Oxford University Press.

Clark, A. 2013a. Expecting the world: Perception, prediction, and the origins of human knowledge. *Journal of Philosophy* 110 (9): 469–496.

Clark, A. 2013b. Whatever next? Predictive brains, situated agents, and the future of cognitive science. *Behavioral and Brain Sciences* 36:181–253.

Clark, A. 2015a. Embodied prediction. In *Open MIND: 7(T)*, ed. T. Metzinger and J. M. Windt. MIND Group.

Clark, A. 2015b. Predicting peace: Reply to Madary. In *Open MIND: 7(R)*, ed. T. Metzinger and J. M. Windt. MIND Group.

Clark, A. 2016. *Surfing Uncertainty: Prediction, Action and the Embodied Mind*. Oxford University Press.

Clark, A., and D. Chalmers. 1998. The extended mind. *Analysis* 58:7–19.

Clark, A., and J. Toribio. 1994. Doing without representing. *Synthese* 101 (3): 401–431.

Clowes, R., and Mendonça, D. 2016. Representation redux: Is there still a useful role for representation to play in the context of embodied, dynamicist and situated theories of mind? *New Ideas in Psychology* 40 (A): 26–47.

Colombetti, G. 2014. *The Feeling Body: Affective Science Meets the Enactive Mind*. MIT Press.

Colombo, M. 2014a. Explaining social norm compliance: A plea for neural representations. *Phenomenology and the Cognitive Sciences* 13 (2): 217–238.

Colombo, M. 2014b. Neural representationalism, the hard problem of content and vitiated verdicts: A reply to Hutto and Myin. *Phenomenology and the Cognitive Sciences* 13 (2): 257–274.

Crane, T. 2009. Is perception a propositional attitude? *Philosophical Quarterly* 59:452–469.

Crane, T. 2014. Human uniqueness and the pursuit of knowledge: A naturalist account. In *Contemporary Philosophical Naturalism and Its Implications*, ed. B. Bashour and H. D. Muller. Routledge.

Craver, C. F. 2007. *Explaining the Brain: Mechanisms and the Mosaic Unity of Neuroscience*. Oxford University Press.

Csibra, G. 2009. Natural pedagogy. *Trends in Cognitive Sciences* 13 (4): 148–153.

Currie, G. 1995. Visual imagery as the simulation of vision. *Mind & Language* 10 (1–2): 25–44.

Currie, G. 2010. *Narratives and Narrators*. Oxford University Press.

Currie, G., and I. Ravenscroft. 2003. *Recreative Minds*. Oxford University Press.

Dale, R., E. Dietrich, and A. Chemero. 2009. Explanatory pluralism in cognitive science. *Cognitive Science* 33 (5): 739–742.

Davids, K., C. Button, and S. Bennett. 2008. *Dynamics of Skill Acquisition*. Human Kinetics.

Davidson, D. 1984. *Inquiries into Truth and Interpretation*. Clarendon Press.

Davidson, D. 1987. Problems in the explanation of action. In *Metaphysics and Morality*, eds. P. Pettit, R. Sylvan, and J. Norman. Blackwell.

Davidson, D. 1999. The emergence of thought. *Erkenntnis* 51 (1): 7–17.

De Brigard, F. 2014. Is memory for remembering? Recollection as a form of episodic hypothetical thinking. *Synthese* 191 (2): 1–31.

Degenaar, J., and E. Myin. 2014. Representation-hunger reconsidered. *Synthese* 191:3639–3648.

De Jesus, P. 2016. From enactive phenomenology to biosemiotic enactivism. *Adaptive Behavior* 24 (2): 130–146.

Dennett, D. C. 1991. *Consciousness Explained*. Little, Brown.

Dennett, D. C. 1995. *Darwin's Dangerous Idea*. Simon and Schuster.

Dennett, D. C. 2013. *Intuition Pumps and Other Tools for Thinking*. Norton.

Deregowski, J. B. 1989. Real space and represented space: Cross-cultural perspectives. *Behavioral and Brain Sciences* 12 (1): 51.

de-Wit, L., D. Alexander, V. Ekroll, and J. Wagemans. 2016. Is neuroimaging measuring information in the brain? *Psychonomic Bulletin & Review*. doi:10.3758/s13423-016-1002-0.

Dijksterhuis, E. 1961. *The Mechanization of the World Picture*. Oxford University Press. Translation by C. Dikshoorn of E. Dijksterhuis, *De Mechanisering van het Wereldbeeld* (Meulenhoff, 1950).

Di Paolo, E. 2005. Autopoiesis, Adaptivity, Teleology, Agency. *Phenomenology and the Cognitive Sciences* 4 (4): 429–452

Di Paolo, E. 2009. Extended life. *Topoi* 28 (1): 9–21.

Di Paolo, E., M. Rohde, and H. De Jaegher. 2010. Horizons for the enactive mind: Values, social interaction, and play. In *Enaction: Towards a New Paradigm for Cognitive Science*, ed. J. Stewart, O. Gapenne, and E. Di Paolo. MIT Press.

Drayson, Z. 2014. The personal/subpersonal distinction. *Philosophy Compass* 9 (5): 338–346.

Dretske, F. 1988. *Explaining Behavior: Reasons in a World of Causes*. MIT Press.

Dreyfus, H. L. 2001. *The primacy of phenomemology over logical analysis*. Reprinted in Dreyfus, *Skillful Coping*.

Dreyfus, H. L. 2002a. Intelligence without representation: Merleau-Ponty's critique of mental representation. *Phenomenology and the Cognitive Sciences* 1:367–383.

Dreyfus, H. L. 2002b. Refocusing the question: Can there be skillful coping without propositional representations or brain representations? *Phenomenology and the Cognitive Sciences* 1 (4): 413–425.

Dreyfus, H. L. 2014. *Skillful Coping: Essays on the Phenomenology of Everyday Perception and Action*, ed. Mark Wrathall. Oxford University Press.

Eisenberg, A. R. 1985. Learning to describe past experiences in conversation. *Discourse Processes* 8:177–204.

Engel, A. K., A. Maye, M. Kurthen, and P. König. 2013. Where's the action? The pragmatic turn in cognitive science. *Trends in Cognitive Sciences* 17 (5): 202–208.

Feynman, R., R. Leighton, and M. Sands. 1963. *The Feynman Lectures on Physics*. Vol. 1. Addison-Wesley.

Fivush, R. 1991. The social construction of personal narratives. *Merrill-Palmer Quarterly* 37:59–82.

Fivush, R. 1997. Event memory in early childhood. In *The Development of Memory in Childhood*, ed. N. Cowan. University College Press.

Fivush, R. 2001. Owning experience: The development of subjective perspective in autobiographical memory. In *The Self in Time: Developmental Perspectives*, ed. C. Moore and K. Lemmon. Erlbaum.

Fivush, R., L. Berlin, J. M. Sales, J. Mennuti-Washburn, and J. Cassidy. 2003. Functions of parent-child reminiscing across negative events. *Memory* 11:179–192.

Fivush, R., J. G. Bohanek, and W. Zaman. 2010. Personal and intergenerational narratives in relation to adolescents' well-being. In *The Development of Autobiographical Reasoning in Adolescence and Beyond: New*

Directions for Child and Adolescent Development 131, ed. T. Habermas. Wiley.

Fivush, R., and F. Fromhoff. 1988. Style and structure in mother-child conversations about the past. *Discourse Processes* 11:337–355.

Fivush, R., T. Habermas, T. E. A. Waters, and W. Zaman. 2011. The making of autobiographical memory: Intersections of culture, narratives and identity. *International Journal of Psychology* 46 (5): 321–345.

Fivush, R., C. Haden, and S. Adam. 1995. Structure and coherence of preschoolers' personal narratives over time: Implications for childhood amnesia. *Journal of Experimental Child Psychology* 60:32–56.

Fivush, R., C. A. Haden, and E. Reese. 1996. Remembering, recounting and reminiscing: The development of autobiographical memory in social context. In *Remembering Our Past: An Overview of Autobiographical Memory*, ed. D. Rubin. Cambridge University Press.

Fivush, R., and J. McDermott Sales. 2006. Coping, attachment, and mother-child narratives of stressful events. *Merrill-Palmer Quarterly* 52 (1): 125–150.

Fivush, R., J. McDermott Sales, and J. G. Bohanek. 2008. Meaning making in mothers' and children's narratives of emotional events. *Memory* 16 (6): 579–594.

Fivush, R., and K. Nelson. 2004. Culture and language in the emergence of autobiographical memory. *Psychological Science* 15:573–577.

Fivush, R., and K. Nelson. 2006. Parent-child reminiscing locates the self in the past. *British Journal of Developmental Psychology* 24:235–251.

Fivush, R., and E. Reese. 1992. The social construction of autobiographical memory. In *Theoretical Perspectives on Autobiographical Memory*, ed. M. A. Conway, D. Rubin, H. Spinnler, and W. Wagenaar. Kluwer Academic.

Fivush, R., and Q. Wang. 2005. Emotion talk in mother-child conversation of the shared past: The effects of culture, gender, and event valence. *Journal of Cognition and Development* 6:489–507.

Flanagan, O. 1991. *The Science of the Mind.* 2nd ed. MIT Press.

Fodor, J. A. 1975. *The Language of Thought.* Harvard University Press.

Fodor, J. A. 1983. *The Modularity of Mind.* MIT Press.

Fodor, J. A. 1987. *Psychosemantics.* MIT Press.

Fodor, J. A. 1990. *A Theory of Content and Other Essays.* MIT Press.

Fodor, J. A. 1991. Fodor's guide to mental representation. In *The Future of Folk Psychology*, ed. J. Greenwood. Cambridge University Press.

Fodor, J. A. 2008a. Against Darwinism. *Mind & Language* 23:1–24.

Fodor, J. A. 2008b. *LOT 2.* MIT Press.

Foglia, L., and R. Grush. 2011. The limitations of a purely enactive (non-representational) account of imagery. *Journal of Consciousness Studies* 18 (5–6): 35–43.

French, L., M. Garry, and K. Mori. 2008. You say tomato? Collaborative remembering leads to more false memories for intimate couples than for strangers. *Memory* 16 (3): 262–273.

Fridland, E. 2014. Skill learning and conceptual thought: Making a way through the wilderness. In *Contemporary Philosophical Naturalism and Its Implications*, ed. B. Bashour and H. D. Muller. Routledge.

Friston, K. J. 2010. The free-energy principle: A unified brain theory? *Nature Reviews: Neuroscience* 11 (2): 127–138.

Friston, K. J., and K. E. Stephan. 2007. Free-energy and the brain. *Synthese* 159:417–458.

Froese, T., and E. A. Di Paolo. 2011. The enactive approach: Theoretical sketches from cell to society. *Pragmatics & Cognition* 19 (1): 1–36.

Froese, T., and T. Ziemke. 2009. Enactive artificial intelligence: Investigating the systemic organization of life and mind. *Artificial Intelligence* 173:466–500.

Gallese, V. 2014. Bodily selves in relation: Embodied simulation as second-person perspective on intersubjectivity. *Philosophical Transactions of the Royal Society B: Biological Sciences* 369 (1644): 20130177.

Gallese, V., and C. Sinigaglia. 2011. What is so special about embodied simulation? *Trends in Cognitive Sciences* 15:512–519.

Gallistel, C. R. 1998. Symbolic processes in the brain: The case of insect navigation. In *An Invitation to Cognitive Science,* Vol. 4: *Methods, Models, and Conceptual Issues*, 2nd ed., ed. D. Scarborough and S. Sternberg. MIT Press.

Garry, M., C. G. Manning, E. F. Loftus, and S. J. Sherman. 1996. Imagination inflation: Imagining a childhood event inflates confidence that it occurred. *Psychonomic Bulletin & Review* 3:208–214.

Gerrans, P. 2014. *The Measure of Madness*. MIT Press.

Gerrans, P., and J. Kennett. 2010. Neurosentimentalism and moral agency. *Mind* 119 (475): 585–614.

Gibson, J. J. 1979. *The Ecological Approach to Visual Perception*. Houghton Mifflin.

Gibson, M. 2004. *From Naming to Saying: The Unity of the Proposition*. Routledge.

Gładziejewski, P. 2015. Explaining cognitive phenomena with internal representations: A mechanistic perspective. *Studies in Logic, Grammar and Rhetoric* 40 (1): 63–90.

Gładziejewski, P. 2016. Predictive coding and representationalism. *Synthese* 193 (2): 559–582.

Godfrey-Smith, P. 2006. Mental representation, naturalism, and teleosemantics. In *Teleosemantics: New Philosophical Essays*, ed. G. Macdonald and D. Papineau. Oxford University Press.

Godfrey-Smith, P. 2009. Representationalism reconsidered. In *Stich and His Critics*, ed. D. Murphy and M. Bishop. Wiley-Blackwell.

Goldie, P. 2004. *On Personality*. Routledge.

Goldie, P. 2012. *The Mess Inside: Narrative, Emotion and the Mind.* Oxford University Press.

Goldman, A., and F. de Vignemont. 2009. Is social cognition embodied? *Trends in Cognitive Sciences* 13:154–159.

Goldman, A. I. 2012. A moderate approach to embodied cognitive science. *Review of Philosophy and Psychology* 3 (1): 71–88.

Goldman, A. I. 2014. The bodily formats approach to embodied cognition. In *Current Controversies in the Philosophy of Mind*, ed. U. Kriegel. Routledge.

Graham, G. 2009. *The Disordered Mind: An Introduction to Philosophy of Mind and Mental Illness.* Routledge.

Grush, R. 1997. The architecture of representation. *Philosophical Psychology* 10:5–23.

Grush, R. 2004. The emulation theory of representation: Motor control, imagery, and perception. *Behavioral and Brain Sciences* 27:377–442.

Gunther, Y. 2003. General introduction. In *Essays on Nonconceptual Content*, ed. Y. Gunther. MIT Press.

Hafting, T., M. Fyhn, S. Molden, M. Moser, and E. Moser. 2005. Microstructure of a spatial map in the entorhinal cortex. *Nature* 436:801–806.

Harley, K., and E. Reese. 1999. Origins of autobiographical memory. *Developmental Psychology* 35:1338–1348.

Harvey, M. 2015. Content in languaging: Why radical enactivism is incompatible with representational theories of language. *Language Sciences* 48:90–129.

Haugeland, J. 1990. The intentionality all-stars. *Philosophical Perspectives* 4:383–427.

Haugeland, J. 1998. Truth and rule-following. In *Having Thought: Essays in the Metaphysics of Mind*, ed. J. Haugeland. Harvard University Press.

Heras-Escribano, M., J. Noble, and M. de Pinedo. 2015. Enactivism, action and normativity: A Wittgensteinian analysis. *Adaptive Behavior* 23:1–14.

Herman, D. 2013. *Storytelling and the Sciences of the Mind*. MIT Press.

Hoerl, C. 2007. Episodic memory, autobiographical memory, narrative: On three key notions in current approaches to memory development. *Philosophical Psychology* 20 (5): 621–640.

Hoerl, C., and T. McCormack. 2005. Joint reminiscing as joint attention to the past. In *Joint Attention: Communication and Other Minds: Issues in Philosophy and Psychology*, ed. N. Eilan, C. Hoerl, T. McCormack, and J. Roessler. Oxford University Press.

Hohwy, J. 2013. *The Predictive Mind*. Oxford University Press.

Hohwy, J. 2014. The self-evidencing brain. *Noûs* 50 (2): 259–285.

Horst, S. 2007. *Beyond Reduction: Philosophy of Mind and Post-Reductionist Philosophy of Science*. Oxford University Press.

Horwich, P. 2012. *Wittgenstein's Metaphilosophy*. Oxford University Press.

Hristovski, R., K. Davids, and D. Araújo. 2009. Information for regulating action in sport: Metastability and emergence of tactical solutions under ecological constraints. In *Perspectives on Cognition and Action in Sport*, ed. D. Araújo, H. Ripoll, and M. Raab. Nova Science Publishers.

Hristovski, R., K. Davids, D. Araújo, and C. Button. 2006. How boxers decide to punch a target: Emergent behaviour in nonlinear dynamical movement systems. *Journal of Sports Science & Medicine* 5:60–73.

Hubbard, E. M., M. Piazza, P. Pinel, and S. Dehaene. 2005. Interactions between number and space in parietal cortex. *Nature Reviews: Neuroscience* 6 (6): 435–448.

Hufendiek, R. 2016. *Embodied Emotions: A Naturalist Approach to a Normative Phenomenon*. Routledge.

Hurley, S., and A. Noë. 2003. Neural plasticity and consciousness: Reply to Block. *Trends in Cognitive Sciences* 7 (8): 342.

Hutto, D. D. 1999. *The Presence of Mind*. Jon Benjamins.

Hutto, D. D. 2005. Knowing what? Radical versus conservative enactivism. *Phenomenology and the Cognitive Sciences* 4 (4): 389–405.

Hutto, D. D. 2006a. Both Bradley and biology: Reply to Rudd. In *Radical Enactivism: Intentionality, Phenomenology and Narrative*, ed. R. Menary. John Benjamins.

Hutto, D. D. 2006b. Unprincipled engagements: Emotional experience, expression and response. In *Radical Enactivism: Intentionality, Phenomenology and Narrative*, ed. R. Menary. John Benjamins.

Hutto, D. D. 2008. *Folk Psychological Narratives: The Sociocultural Basis of Understanding Reasons*. MIT Press.

Hutto, D. D. 2011. Elementary mind minding, enactivist-style. In *Joint Attention: New Developments in Philosophy, Psychology, and Neuroscience*, ed. A. Seemann. MIT Press.

Hutto, D. D. 2013a. Enactivism: From a Wittgensteinian point of view. *American Philosophical Quarterly* 50 (3): 281–302.

Hutto, D. D. 2013b. Exorcising action oriented representations: Ridding cognitive science of its Nazgûl. *Adaptive Behavior* 21 (1): 142–150.

Hutto, D. D. 2013c. Fictionalism about folk psychology. *Monist* 96 (4): 585–607.

Hutto, D. D. 2014. Contentless perceiving: The very idea. In *Wittgenstein and Perception*, ed. M. Campbell and M. O'Sullivan. Routledge.

Hutto, D. D. 2015a. Basic social cognition without mindreading: Minding minds without attributing contents. *Synthese*: 1–20. doi:10.1007/s11229-015-0831-0.

Hutto, D. D. 2015b. Overly enactive imagination? Radically re-imagining imagining. *Southern Journal of Philosophy* 53 (S1): 68–89.

Hutto, D. D. 2016. Narrative self-shaping: A modest proposal. *Phenomenology and the Cognitive Sciences* 15:21–41.

Hutto, D. D. In press. Memory and narrativity. In *Handbook of Philosophy of Memory*, ed. S. Bernecker and K. Michaelian. Routledge.

Hutto, D. D., M. D. Kirchhoff, and E. Myin. 2014. Extensive enactivism: Why keep it all in? *Frontiers in Human Neuroscience* 8:706.

Hutto, D. D., and E. Myin. 2013. *Radicalizing Enactivism: Basic Minds without Content*. MIT Press.

Hutto, D. D., and E. Myin. 2014. Neural representations not needed: No more pleas, please. *Phenomenology and the Cognitive Sciences* 13 (2): 241–256.

Hutto, D. D., and R. Sánchez-García. 2015. Choking RECtified: Embodied Expertise beyond Dreyfus. *Phenomenology and the Cognitive Sciences* 14 (2): 309–331.

Hutto, D. D., and G. Satne. 2015. The natural origins of content. *Philosophia* 43 (3): 521–536.

Intraub, H., and M. Richardson. 1989. Wide-angle memories of close-up scenes. *Journal of Experimental Psychology: Learning, Memory, and Cognition* 15:179–187.

Jackson, F., and P. Pettit. 1993. Some content is narrow. In *Mental Causation*, ed. J. Heil, and A. Mele, 259–282. Oxford University Press.

Jacob, P. 2014. Intentionality. In *The Stanford Encyclopedia of Philosophy*, Winter 2014 edition, ed. E. N. Zalta. Metaphysics Research Lab/Stanford University.

Janssen, S. M. J., A. G. Chessa, and J. M. J. Murre. 2006. Memory for time: How people date events. *Memory & Cognition* 34:138–147.

Kamitani, Y., and F. Tong. 2005a. Decoding seen and attended motion directions from activity in the human visual cortex. *Current Biology* 16 (11): 1096–1102.

Kamitani, Y., and F. Tong. 2005b. Decoding the visual and subjective contents of the human brain. *Nature Reviews: Neuroscience* 8 (5): 679–685.

Kandel, E. 2001. The molecular biology of memory storage: A dialogue between genes and synapses. *Science* 294:1030–1038.

Kandel, E. 2009. The biology of memory: A forty-year perspective. *Journal of Neuroscience* 29 (41): 12748–12756.

Kaplan, D. M. 2015. Moving parts: The natural alliance between dynamical and mechanistic modeling approaches. *Biology & Philosophy* 30 (6): 757–786.

Kelso, S. 1995. *Dynamic Patterns*. MIT Press.

Keysers, C., J. H. Kaas, and V. Gazzola. 2010. Somatosensation in social perception. *Nature Reviews: Neuroscience* 11:417–428.

Khalidi, M. A. 2007. Innate cognitive capacities. *Mind & Language* 22 (1): 92–115.

Kiverstein, J., and E. Rietveld. 2015. The primacy of skilled intentionality: On Hutto and Satne's The Natural Origins of Content. *Philosophia* 43:701–721.

Korbak, T. 2015. Scaffolded minds and the evolution of content in signaling pathways. *Studies in Logic, Grammar and Rhetoric* 41 (54): 89–103.

Kriegel, U. 2011. *The Sources of Intentionality*. Oxford University Press.

Langland-Hassan, P. 2015. Imaginative attitudes. *Philosophy and Phenomenological Research* 90 (3): 664–686.

Lavelle, J. S. 2012. Two challenges to Hutto's enactive account of pre-linguistic social cognition. *Philosophia* 40 (3): 459–472.

Lindsay, D. S., L. Hagen, J. D. Read, K. A. Wade, and M. Garry. 2004. True photographs and false memories. *Psychological Science* 15:149–154.

Lindsay, P, and D. A. Norman. *Human Information Processing: An Introduction to Psychology*. 2nd ed. Academic Press.

Loftus, E. F. 1996. *Eyewitness Testimony*. 2nd ed. Harvard University Press.

Loftus, E. F. 2005. Planting misinformation in the human mind: A 30-year investigation of the malleability of memory. *Learning & Memory (Cold Spring Harbor, NY)* 12 (4): 361–366.

Loftus, E. F., D. G. Miller, and H. J. Burns. 1978. Semantic integration of verbal information into a visual memory. *Journal of Experimental Psychology: Human Learning and Memory* 4 (1): 19–31.

Loftus, E. F., and J. E. Pickrell. 1995. The formation of false memories. *Psychiatric Annals* 25 (12): 720–725.

Lungarella, M., and O. Sporns. 2005. Information self-structuring: Key principle for learning and development. In *Proceedings 2005 IEEE International Conference on Development and Learning*, 25–30.

Lupyan, G., and A. Clark. 2015. Words and the world: Predictive coding and the language-perception-cognition interface. *Current Directions in Psychological Science* 24 (4): 279–284.

Machery, E. 2009. *Doing without Concepts*. Oxford University Press.

Malafouris, M. 2013. *How Things Shape the Mind: A Theory of Material Engagement*. MIT Press.

Mandik, P. 2005. *Cognition and the Brain: The Philosophy and Neuroscience Movement*. Ed. A. Brook and K. Akins. Cambridge University Press.

Marr, D. 1982. *Vision: A Computational Investigation into the Human Representation and Processing of Visual Information*. W.H. Freeman and Co.

Matthen, M. 2014. Debunking enactivism: A critical notice of Hutto and Myin's Radicalizing Enactivism. *Canadian Journal of Philosophy* 44 (1): 118–128.

McCauley, R., and J. Henrich. 2006. Susceptibility to the Müller-Lyer illusion, theory-neutral observation, and the diachronic penetrability of the visual input system. *Philosophical Psychology* 19 (1): 1–23.

McDermott Sales, J., R. Fivush, J. Parker, and L. Bahrick. 2005. Stressing memory: Long-term relations among children's stress, recall and psychological outcome following Hurricane Andrew. *Journal of Cognition and Development* 6 (4): 529–545.

McDowell, J. 1994. *Mind and World*. Harvard University Press.

McDowell, J. 1998. *Mind, Value, and Reality*. Harvard University Press.

McGinn, C. 1989. *Mental Content*. Blackwell.

McGinn, C. 2004. *Mindsight: Image, Dream, Meaning*. Harvard University Press.

McIsaac, H. K., and E. Eich. 2002. Vantage point in episodic memory. *Psychonomic Bulletin & Review* 9 (1): 146–150.

Medina, J. 2013. An enactivist approach to the imagination: Embodied enactments and fictional emotions. *American Philosophical Quarterly* 50 (3): 317–335.

Menary, R. 2007. *Cognitive Integration: Mind and Cognition Unbounded*. Palgrave.

Menary, R. 2009. Intentionality, cognitive integration and the continuity thesis. *Topoi* 28 (1): 31–43.

Menary, R. 2015a. Mathematical cognition—A case of enculturation. In *Open MIND: 25 (T)*, ed. T. Metzinger and J. M. Windt. MIND Group.

Menary, R. 2015b. What now? Predictive coding and enculturation. In *Open MIND: 25 (T)*, ed. T. Metzinger and J. M. Windt. MIND Group.

Merritt, M. 2015. Dismantling standard cognitive science: It's time the dog has its day. *Biology & Philosophy* 30 (6): 811–829.

Michaelian, K. 2011. Is memory a natural kind? *Memory Studies* 4 (2): 170–189.

Michaelian, K. 2016. *Mental Time Travel: Episodic Memory and Our Knowledge of the Personal Past.* MIT Press.

Miłkowski, M. 2015. The hard problem of content: Solved (long ago). *Studies in Logic, Grammar and Rhetoric* 41 (1): 73–88.

Miller, P. J., R. Potts, H. Fung, L. Hoogstra, and J. Mintz. 1990. Narrative practices and the social construction of the self in childhood. *American Ethnologist* 17:292–311.

Millikan, R. G. 1984. *Language, Thought, and Other Biological Categories.* MIT Press.

Millikan, R. G. 1993. *White Queen Psychology and Other Essays for Alice.* MIT Press.

Millikan, R. G. 2004. *Varieties of Meaning: The 2002 Jean Nicod Lectures.* MIT Press.

Millikan, R. G. 2005. *Language: A Biological Model.* Oxford University Press.

Mole, C., and J. Zhao. 2015. Vision and abstraction: An empirical refutation of Nico Orlandi's non-cognitivism. *Philosophical Psychology* 29 (3): 365–373.

Moser, E., and M. Moser. 2008. A metric for space. *Hippocampus* 18:1142–1156.

Moyal-Sharrock, D. 2009. Wittgenstein and the memory debate. *New Ideas in Psychology* 27:213–227.

Muller, H. D. 2014. Naturalism and intentionality. In *Contemporary Philosophical Naturalism and Its Implications*, ed. B. Bashour and H. D. Muller. London: Routledge.

Myin, E., and J. Degenaar. 2014. Enactive vision. In *The Routledge Handbook of Embodied Cognition*, ed. L. Shapiro. London: Routledge.

Myin, E., and D. D. Hutto. 2015. REC: Just radical enough. *Studies in Logic, Grammar and Rhetoric* 41 (1): 61–71.

Myin, E., and J. Veldeman. 2011. Externalism, mind and art. In *Situated Aesthetics: Art beyond the Skin*, ed. R. Manzotti. Imprint Academic.

Nanay, B. 2014. Empirical problems with anti-representationalism. In *Does Perception Have Content?*, ed. B. Brogaard. Oxford University Press.

Nelson, K. 1988. The ontogeny of memory for real events. In *Remembering Reconsidered: Ecological and Traditional Approaches to the Study of Memory*, ed. U. Neisser and E. Winograd. Cambridge University Press.

Nelson, K. 1993. The psychological and social origins of autobiographical memory. *Psychological Science* 4:7–14.

Nelson, K. 1996. *Language in Cognitive Development: Emergence of the Mediated Mind*. Cambridge University Press.

Nelson, K. 2003. Narrative and the emergence of a consciousness of self. In *Narrative and Consciousness*, ed. G. D. Fireman, T. E. J. McVay, and O. Flanagan. Oxford University Press.

Nelson, K. 2007. *Young Minds in Social Worlds: Experience, Meaning, and Memory*. Harvard University Press.

Nelson, K., and R. Fivush. 2004. The emergence of autobiographical memory: A social cultural developmental theory. *Psychological Review* 111:486–511.

Neter, J., and J. Waksberg. 1964. A study of response errors in expenditures data from household interviews. *American Statistical Association Journal* 59:18–55.

Newell, K. M. 1986. Constraints on the development of coordination. In *Motor Development in Children: Aspects of Coordination and Control*, ed. M. G. Wade and H. T. A. Whiting. Nijhoff.

Nichols, S., ed. 2006. *The Architecture of the Imagination: New Essays on Pretence, Possibility, and Fiction*. Oxford University Press.

Nigro, G., and U. Neisser. 1983. Point of view in personal memories. *Cognitive Psychology* 15:467–482.

Noë, A. 2004. *Action in Perception*. MIT Press.

Noë, A. 2009. *Out of Our Heads*. Hill and Wang.

Noë, A. 2012. *Varieties of Presence*. Harvard University Press.

O'Brien, G., and J. Opie. 2009. The role of representation in computation. *Cognitive Processing* 10 (1): 53–62.

O'Brien, G., and J. Opie. 2015. Intentionality lite or analog content? *Philosophia* 43 (3): 723–730.

O'Keefe, J. 1976. Place units in the hippocampus of the freely moving rat. *Experimental Neurology* 51 (1): 78–109.

O'Keefe, J., and J. Dostrovsky. 1971. The hippocampus as a spatial map: Preliminary evidence from unit activity in the freely-moving rat. *Journal of Brain Research* 34 (1): 171–175.

O'Keefe, J., and L. Nadel. 1978. *The Hippocampus as a Cognitive Map*. Oxford University Press.

Ólafsdóttir, F., C. Barry, A. Saleem, D. Hassabis, and H. Spiers. 2015. Hippocampal place cells construct reward related sequences through unexplored space. *eLife* 4:e06063.

O'Regan, J. K., and A. Noë. 2001. A sensorimotor account of vision and visual consciousness. *Behavioral and Brain Sciences* 24:939–1031.

Orlandi, N. 2014. *The Innocent Eye: Why Vision Is Not a Cognitive Process*. Oxford University Press.

Papineau, D. 1987. *Reality and Representation*. Oxford University Press.

Papineau, D. 2014. The poverty of conceptual analysis. In *Philosophical Methodology: The Armchair or the Laboratory?*, ed. M. C. Haug. Routledge.

Pattee, H. 1985. Universal principles of measurement and language functions in evolving systems. In *Complexity of Language and Life: Mathematical Approaches*, ed. J. Casti and A. Karlqvist. Springer-Verlag.

Penn, D. C., K. J. Holyoak, and D. J. Povinelli. 2008. Darwin's mistake: Explaining the discontinuity between human and nonhuman minds. *Behavioral and Brain Sciences* 31:109–178.

Pezdek, K., I. Blandon-Gitlin, and P. Gabbay. 2006. Imagination and memory: Does imagining implausible events lead to false autobiographical memories? *Psychonomic Bulletin & Review* 13 (5): 764–769.

Piccinini, G. 2008. Computation without representation. *Philosophical Studies* 137 (2): 205–241.

Piccinini, G. 2015. *Physical Computation: A Mechanistic Account.* Oxford University Press.

Pietroski, M. 1992. Intentionality and teleological error. *Pacific Philosophical Quarterly* 73:267–282.

Pillemer, D. B., and S. H. White. 1989. Childhood events recalled by children and adults. In *Advances in Child Development and Behavior*, ed. H. W. Reese. Academic Press.

Price, H. 2013. *Expressivism, Pragmatism and Representationalism.* Cambridge University Press.

Prinz, J. 2002. *Furnishing the Mind.* MIT Press.

Proffitt, D. R. 2008. An action-specific approach to spatial perception. In *Embodiment, Ego-Space, and Action*, ed. R. L. Katzky, B. MacWhinney, and M. Behrmann. Psychology Press.

Pulvermuller, F. 2005. Brain mechanisms linking language and action. *Nature Reviews: Neuroscience* 6:576–582.

Putnam, H. 1988. *Representation and Reality.* MIT Press.

Putnam, H. 1992. *Renewing Philosophy.* Harvard University Press.

Rakoczy, H., F. Warneken, and M. Tomasello. 2008. The sources of normativity: Young children's awareness of the normative structure of games. *Developmental Psychology* 44 (3): 875–881.

References

Rakoczy, H., F. Warneken, and M. Tomasello. 2009. Young children's selective learning of rule games from reliable and unreliable models. *Cognitive Development* 24:61–69.

Ramsey, W. M. 2007. *Representation Reconsidered*. Cambridge University Press.

Ramsey, W. M. 2014. Must cognition be representational? *Synthese*. doi:10.1007/s11229-014-0644-6.

Ramstead, M. J. D., S. P. L. Veissière, and L. J. Kirmayer. 2016. Cultural affordances: Scaffolding local worlds through shared intentionality and regimes of attention. *Frontiers in Psychology* 7:1090. doi:10.3389/fpsyg.2016.01090.

Reddy, L., N. Tsuchiya, and T. Serre. 2010. Reading the mind's eye: Decoding category information during mental imagery. *NeuroImage* 50 (2): 818–825.

Reese, E., and R. Fivush. 2008. Collective memory across the lifespan. *Memory* 16:201–212.

Rescorla, M. 2012a. Are computational transitions sensitive to semantics? *Australasian Journal of Philosophy* 90 (4): 703–721.

Rescorla, M. 2012b. Millikan on honeybee navigation and communication. In *Millikan and Her Critics*, ed. D. Ryder, J. Kingsbury, and K. Williford. Wiley.

Rescorla, M. 2014. The causal relevance of content to computation. *Philosophy and Phenomenological Research* 88 (1): 173–208.

Rescorla, M. 2016. Bayesian sensorimotor psychology. *Mind & Language* 31 (1): 3–36.

Rey, G. 2002. Problems with Dreyfus' Dialectic. *Phenomenology and the Cognitive Sciences* 1 (4): 403–408.

Rey, G. 2015. Representation. In *The Bloomsbury Companion to the Philosophy of Mind*, ed. J. Garvey. Bloomsbury.

Richner, E. S., and A. Nicolopoulou. 2001. The narrative construction of differing conceptions of the person in the development of young children's social understanding. *Early Education and Development* 12:393–432.

Rickles, D., P. Hawe, and A. Shiell. 2007. A simple guide to chaos and complexity. *Journal of Epidemiology and Community Health* 61:933–937.

Rietveld, E. 2008. Situated normativity: The normative aspect of embodied cognition in unreflective action. *Mind* 117 (468): 973–1001.

Rietveld, E., and J. Kiverstein. 2014. A rich landscape of affordances. *Ecological Psychology* 26 (4): 325–352.

Riley, M. A., and J. G. Holden. 2012. Dynamics of cognition. *Wiley Interdisciplinary Reviews: Cognitive Science* 3 (6): 593–606.

Ritchie, J. 2008. *Understanding Naturalism*. Acumen.

Rizzolatti, G., L. Fadiga, V. Gallese, and L. Fogassi. 1996. Premotor cortex and the recognition of motor actions. *Brain Research/Cognitive Brain Research* 3:131–141.

Rizzolatti, G., and C. Sinigaglia. 2010. The functional role of the parieto-frontal mirror circuit: Interpretations and misinterpretations. *Nature Reviews: Neuroscience* 11:264–274.

Robinson, J. A., and K. L. Swanson. 1993. Field and observer modes of remembering. *Memory* 1 (3): 169–184.

Roediger, H. L. 1980. Memory metaphors in cognitive psychology. *Memory & Cognition* 8 (3): 231–246.

Roediger, H. L., and K. B. McDermott. 1995. Creating false memories: Remembering words not presented in lists. *Journal of Experimental Psychology: Learning, Memory, and Cognition* 21:803–814.

Rosenberg, A. 2013. How Jerry Fodor slid down the slippery slope to anti-Darwinism, and how we can avoid the same fate. *European Journal of Philosophy of Science* 3:1–17.

Rosenberg, A. 2014a. Can naturalism save the humanities? In *Philosophical Methodology: The Armchair or the Laboratory?*, ed. M. C. Haug. Routledge.

Rosenberg, A. 2014b. Disenchanted naturalism. In *Contemporary Philosophical Naturalism and Its Implications*, ed. B. Bashour and H. D. Muller. Routledge.

Rosenberg, A. 2014c. Why I am a naturalist. In *Philosophical Methodology: The Armchair or the Laboratory?*, ed. M. C. Haug. Routledge.

Rosenberg, A. 2015. The genealogy of content or the future of an illusion. *Philosophia* 43:537–547.

Rowlands, M. 2009. The extended mind. *Zygon* 44:628–641.

Rowlands, M. 2015a. Arguing about representation. *Synthese*. doi:10.1007/s11229-014-0646-4.

Rowlands, M. 2015b. Hard problems of intentionality. *Philosophia* 43 (3): 741–746.

Roy, J.-M. 2015. Anti-Cartesianism and anti-Brentanism: The problem of anti-representationalist intentionalism. *Southern Journal of Philosophy* 53:90–125.

Rupert, R. 2009. *Cognitive Systems and the Extended Mind*. Oxford University Press.

Rupert, R. 2010. Extended cognition and the priority of cognitive systems. *Cognitive Systems Research* 11:343–356.

Sainsbury, R. M. 2010. *Fiction and Fictionalism*. Routledge.

Satne, G. 2014. Interaction and self-correction. In "Towards an Embodied Science of Intersubjectivity," ed. E. Di Paolo and H. De Jaegher, special issue, *Frontiers in Psychology* 5:798. doi:10.3389/fpsyg.2014.00798.

Schachtel, E. 1947. On memory and childhood amnesia. *Psychiatry* 10:1–26.

Schacter, D. L., and D. R. Addis. 2009. On the nature of medial temporal lobe contributions to the constructive simulation of future events. *Philosophical Transactions of the Royal Society B: Biological Sciences* 364:1245–1253.

Schacter, D. L., D. R. Addis, and R. L. Buckner. 2007. Remembering the past to imagine the future: The prospective brain. *Nature Reviews: Neuroscience* 8:657–661.

Schacter, D. L., D. R. Addis, and R. L. Buckner. 2008. Episodic simulation of future events: Concepts, data, and applications. *Annals of the New York Academy of Sciences* 1124:39–60.

Schacter, D. L., D. R. Addis, D. Hassabis, V. C. Martin, R. N. Spreng, and K. K. Szpunar. 2012. The future of memory: Remembering, imagining, and the brain. *Neuron* 76 (4).

Schacter, D. L., and E. Tulving. 1994. What are the memory systems of 1994? In *Memory Systems*, ed. D. L. Schacter and E. Tulving. MIT Press.

Schechtman, M. 1994. The truth about memory. *Philosophical Psychology* 7:3–18.

Schechtman, M. 1996. *The Constitution of Selves*. Cornell University Press.

Schechtman, M. 2007. Stories, lives, and basic survival: A refinement and defense of the narrative view. In *Narrative and Understanding Persons*, ed. D. D. Hutto. Cambridge University Press.

Schechtman, M. 2011. The narrative self. In *The Oxford Handbook of the Self*, ed. S. Gallagher. Oxford University Press.

Seager, W. 2000. *Theories of Consciousness*. Routledge.

Searle, J. 1983. *Intentionality: An Essay in the Philosophy of Mind*. Cambridge University Press.

Searle, J. R. 2011. Wittgenstein and the background. *American Philosophical Quarterly* 48 (2): 119–128.

Segall, M., D. Campbell, and M. J. Herskovits. 1966. *The Influence of Culture on Visual Perception*. Bobbs-Merrill.

Shams, L. 2012. Early integration and bayesian causal inference in multisensory perception. In *The Neural Bases of Multisensory Processes*, ed. M. M. Murray and M. T. Wallace. CRC Press/Taylor & Francis.

Shams, L., Y. Kamitani, and S. Shimojo. 2000. What you see is what you hear. *Nature* 408:788.

Shapiro, L. 2014a. Radicalizing Enactivism: Basic Minds without Content, by Daniel D. Hutto and Erik Myin. [Review.] *Mind* 123 (489): 213–220.

Shapiro, L. 2014b. When is cognition embodied? In *Current Controversies in Philosophy of Mind*, ed. U. Kriegel. Routledge.

Shea, N. 2013. Naturalising representational content. *Philosophy Compass* 8 (5): 496–509.

Shepard, R. N., and L. A. Cooper. 1982. *Mental Images and Their Transformations*. MIT Press.

Siegel, S. 2014. Affordances and the contents of perception. In *Does Perception Have Content?*, ed. B. Brogaard. Oxford University Press.

Silberstein, M., and A. Chemero. 2013. Constraints on localization and decomposition as explanatory strategies in the biological sciences. *Philosophy of Science* 80 (5): 958–970.

Silverman, D. 2013. Sensorimotor enactivism and temporal experience. *Adaptive Behavior* 21 (3): 151–158.

Slotnick, S., W. Thompson, and S. M. Kosslyn. 2005. Visual mental imagery induces retinotopically organized activation of early visual areas. *Cerebral Cortex* 15:1570–1583.

Spaulding, S. 2011. A critique of embodied simulation. *Review of Philosophy and Psychology* 2 (3): 579–599.

Sprevak, M. 2013. Fictionalism about neural representations. *Monist* 96:539–560.

Sterelny, K. 2012. *The Evolved Apprentice*. MIT Press.

Sterelny, K. 2015. Content, control and display: The natural origins of content. *Philosophia* 43:549–564.

Stern, D. G. 1991. Models of memory: Wittgenstein and cognitive science. *Philosophical Psychology* 4 (2): 203–218.

Steward, H. 2016. Making the agent reappear: How processes might help. In *Time and the Philosophy of Action*, ed. R. Altshuler and M. J. Sigrist. Routledge.

Stewart, J., O. Gapenne, and E. Di Paolo, eds. 2010. *Enaction: Toward a New Paradigm for Cognitive Science*. MIT Press.

Stich, S. 1983. *From Folk Psychology to Cognitive Science*. MIT Press.

Stich, S. 1990. *The Fragmentation of Reason: Preface to a Pragmatic Theory of Cognitive Evaluation*. MIT Press.

Strawson, G. 1994. *Mental Reality*. MIT Press.

Strawson, G. 2004. Against narrativity. *Ratio* 17 (4): 428–452.

Sutton, J. 2007. Integrating the philosophy and psychology of memory: Two case studies. In *Cartographies of the Mind*, ed. M. Marraffa, M. De Caro, and F. Ferretti. Springer-Verlag.

Sutton, J. 2010. Observer perspective and acentred memory: Some puzzles about point of view in personal memory. *Philosophical Studies* 148:27–37.

Sutton, J. 2015. Remembering as public practice: Wittgenstein, memory, and Distributed Cognitive Ecologies. In *Mind, Language, and Action: Proceedings of the 36th Wittgenstein Symposium*, ed. D. Moyal-Sharrock, V. Munz, and A. Coliva. De Gruyter.

Sutton, J., and D. McIlwain. 2015. Breadth and depth of knowledge in expert versus novice athletes. In *The Routledge Handbook of Sport Expertise*, ed. J. Baker and D. Farrow. Routledge.

Sutton, J., D. McIlwain, W. Christensen, and A. Geeves. 2011. Applying intelligence to the reflexes: Embodied skills and habits between Dreyfus and Descartes. *Journal of the British Society for Phenomenology* 42 (1): 78–103.

Sutton, J., and K. Williamson. 2014. Embodied remembering. In *The Routledge Handbook of Embodied Cognition*, ed. L. Shapiro. Routledge.

Szpunar, K. K., J. M. Watson, and K. B. McDermott. 2007. Neural substrates of envisioning the future. *Proceedings of the National Academy of Sciences of the United States of America* 104:642–647.

Tallis, R. 2003. *The Hand: A Philosophical Inquiry in Human Being*. Edinburgh University Press.

Thagard, P. 1992. *Conceptual Revolutions*. Princeton University Press.

Thelen, E., and L. B. Smith. 2001. The dynamics of embodiment: A field theory of infant perseverative reaching. *Behavioral and Brain Sciences* 24 (1): 1–86.

Thompson, C. P., J. J. Skowronski, S. F. Larsen, and A. L. Betz. 1996. *Autobiographical Memory: Remembering What and Remembering When*. Erlbaum.

Thompson, E. 2007. *Mind in Life: Biology, Phenomenology, and the Sciences of the Mind*. Harvard University Press.

Thompson, E., and M. Stapleton. 2009. Making sense of sense-making: Reflections on enactive and extended mind theories. *Topoi* 28 (1): 23–30.

Tomasello, M. 1999. *The Cultural Origins of Human Cognition*. Harvard University Press.

Tonneau, F. 2011/2012. Metaphor and truth: A review of *Representation Reconsidered* by W. M. Ramsey. *Behavior and Philosophy* 39/40:331–343.

Travis, C. 2004. The silence of the senses. *Mind* 113 (449): 57–94.

Tulving, E. 1972. Episodic and semantic memory. In *Organization of Memory*, ed. E. Tulving and W. Donaldson. Academic Press.

van den Herik, J. 2014. Why Radical Enactivism is not radical enough: A case for really Radical Enactivism. Unpublished master's thesis. Faculty of Philosophy, University of Rotterdam.

van Dijk, L., R. Withagen, and R. M. Bongers. 2015. Information without content: A Gibsonian reply to enactivists' worries. *Cognition* 134:210–214.

van Leeuwen, C., I. M. Verstijnen, and P. Hekkert. 1999. Common unconscious dynamics underlie uncommon conscious effect: A case study in the iterative nature of perception and creation. In *Modeling Consciousness across the Disciplines*, ed. J. S. Jordan. University Press of America.

Varela, F., E. Thompson, and E. Rosch. 1991. *The Embodied Mind: Cognitive Science and Human Experience*. MIT Press.

Wade, K. A., M. Garry, D. Read, and D. S. Lindsay. 2002. A picture is worth a thousand lies: Using false photographs to create false childhood memories. *Psychonomic Bulletin & Review* 9 (3): 597–603.

Weber, A., and F. Varela. 2002. Life after Kant: Natural purposes and the autopoietic foundations of biological individuality. *Phenomenology and the Cognitive Sciences* 1:97–125.

Wheeler, M. 2010. In defense of extended functionalism. In *The Extended Mind*, ed. R. Menary. MIT Press.

Wheeler, M. 2015. The revolution will not be optimised: Radical enactivism, extended functionalism and the extensive mind. *Topoi*: 1–16. doi:10.1007/s11245-015-9356-x.

Williamson, T. 2014a. The unclarity of naturalism. In *Philosophical Methodology: Armchair or Laboratory?*, ed. M. C. Haug. Routledge.

Williamson, T. 2014b. What is naturalism? In *Philosophical Methodology: Armchair or Laboratory?*, ed. M. C. Haug. Routledge.

Wilson, R. A., and L. Foglia. 2016. Embodied Cognition. In *The Stanford Encyclopedia of Philosophy*, Spring 2016 edition, ed. E. N. Zalta. Metaphysics Research Lab/Stanford University.

Wittgenstein, L. 1953. *Philosophical Investigations*. Blackwell.

Yablo, S. 1993. Is conceivability a guide to possibility? *Philosophy and Phenomenological Research* 53 (1): 1–42.

Zednik, C. 2011. The nature of dynamical explanation. *Philosophy of Science* 78 (2): 238–263.

Index

Aboutness, 96–100, 114.
  See also Intentionality, Ur-intentionality
Abramova, Katja, 116–117, 125
Absent, 161–162, 257n5.
  See also Abstract; Representation-hunger
Abstract, 33–35, 64, 158, 161–162. *See also* Absent; Representation-hunger
Accuracy conditions, 11, 45, 100, 147, 152–153, 160, 183, 279n19
Action-oriented, 57, 63–64
Adam, Salimah, 212
Adams, Fred, 282n5
Addis, Donna, 217
Affordance, 78, 82–88, 119
Aizawa, Kenneth, 4–7, 38–40, 237, 258n9, 282n5
Akins, Kathleen, 5
Alksnis, Nikolas, 117, 122–123
Allsmith, Adrian, 5
Anderson, Michael, 185, 217

Animal cognition, 28, 90, 117, 127, 131, 135, 143, 202, 266n19, 268–269n11. *See also* Continuity; Navigation
Anticipation, xiv, xvi, 58–59, 70, 72, 152–153, 196, 224, 253
Antirealism, 45–47, 70, 259n13. *See also* Fictionalism
Araújo, Duarte, 118
Artifacts, 26, 175, 193–194, 204, 207–208
Attention, 25, 33, 59, 143, 189
  anchors for, 117
  shared or joint, 141, 209, 254
Attitude
  cognitive, 11–12, 70, 221
  content-involving, xxi, 140, 145, 171, 176, 192–195
  contentless, 102, 138, 188
  expressive, 140, 143–144
  imaginative, xxi, 182, 189–193, 274n6, 275n12
  intentional, xii, 95–97, 102, 138–144, 195

Attitude (cont.)
propositional, 95, 100, 102, 182, 262n3, 263n6, 272n13
Autonomy, 76–78
Autonomy thesis, 167–170, 271–272n13, 272n14
Autopoiesis, 76–78
Autopoietic-adaptive enactivism (AAE), 75–81, 88

Barandiaran, Xabier, 29
Barnier, Amanda, 213, 220, 229–231
Bar-on, Dorit, 131–132, 136, 143, 268n9, 269n13
Barth, Christian, 127
Bartlett, Frederic, 204, 277n13
Basic minds, vs. nonbasic minds, xii, xix, 13, 89–92, 122–123, 128–130, 134–137, 170–176
Bayesian
activity, 155
approaches, 56–57, 70, 150–151
epistemology, 261n10
inferences, 153
laws, 152
modelling, 270n3
norms, 152, 261n8, 10, 270n3
predictions, 58
rules, 151, 270
Bechtel, William, 4, 241–243, 245
Beer, Randall, 39, 251
Behavior, 4, 14–18, 24, 33–35, 39–40, 107–111, 116–117, 132, 141–144, 158, 162, 186, 196, 235, 241, 243, 245, 247, 250, 269. *See also* Radically Enactive Account of Behavior
Behaviorism, 141–144, 268n11. *See also* Radically Enactive Account of Behavior
Bennett, Simon, 23
Bermúdez, José, 274n10
Biological
function, 45, 104–109
normativity, 76–80
Biosemantics, 44, 238–239. *See also* Teleosemantics
Biosemiotics, 80, 142
Blandon–Gitlin, Iris, 278n16
Bohanek, Jennifer, 211, 229
Bongers, Raoul, 86–88
Bootstrap, 68–70, 154–156, 261n8
Brain. *See also* Hominin, brain; Reuse
as analogous to scientist, xiv, 64–71, 74, 152, 160
changes, xviii, 26–28, 165, 235, 239
coding, 66
and computation, 4–7, 60, 141, 252, 270n6
and contentful representation, 5, 61–64, 66, 71–74, 107, 110, 153, 160, 195, 233–235, 237–239, 246
and contentless information, 74, 238, 245
decoding, 65–66, 74, 185

and homunculus, 241
implants, 6
and information processing, xiv, xxii, 4, 39, 65, 195, 233–235, 237–239, 246
as interactive and dynamic, xiv, 25
as making inferences, 62, 64–74, 153, 160, 260n4
models of, 38, 258n7
models or maps in, 8, 59, 61, 74, 81, 153–154, 233, 240–245
as part of a larger system, 7, 25, 40, 81, 101, 196, 238, 251
predictive, 57–61, 66, 71–72, 81, 180, 259n1
reading (*see* Brain, decoding)
role in cognition, 4, 38, 40, 64, 72–74, 101, 237–245
secluded, 73
Brainerd, Charles, 222
Branquinho, João, 3
Brentano, Franz, 93–99, 114, 263n2, 267n4
Brogaard, Berit, 11, 149–150
Brook, Andrew, 255n2
Bruineberg, Jelle, 10, 70, 117–118
Buckner, Randy, 217
Burge, Tyler, 151, 270n5
Burns, Helen, 224–225
Button, Chris, 23
Byrge, Lisa, 70

Campbell, Donald, 175
Campbell, Douglas, 148–149
Campbell, John, 170
Campbell, Sue, 226, 228, 230–231
Cao, Rosa, 44, 130, 238–239, 267n6
Cappuccio, Massimiliano, 82, 255n1
Carpentered-environment hypothesis, 175
Carruthers, Peter, 18, 108, 142
Casey, Edward, 205
Cash, Mason, 106
Chalmers, David, 6, 48
Chemero, Anthony, 10, 24, 40, 84–86, 250–251
Chessa, Antonio, 224
Chomsky, Noam, 255–256n2
Chow, Jia Yi, 23, 118
Churchland, Paul, 258n8
Claim-making practices, 145–146, 231
Clapin, Hugh, 124
Clark, Andy, 6–7, 25, 34, 56–74, 81–85, 146, 161, 172, 175, 180, 185, 237, 253, 257n5, 259nn1–2, 260nn3,6, 261n8
Clowes, Rob, 129, 177, 179–180, 272n1
Cognitive niche, 134, 137, 146
Cognitive science, xiv–xvi, 1, 4, 5, 14–15, 30, 36–38, 46, 52, 82–85, 111, 150, 152, 166, 177, 223, 247, 250–251, 253, 255n1, 269n1, 270n6, 272n15
successes of, 33, 48–50, 55

Cognitivism, 3–9, 30–33, 39–40, 46–53, 59–70, 81–86, 88, 93, 109, 151, 154–156, 167, 169, 195, 236, 239, 255n1, 257n3, 260nn3,5, 269n15, 273n3
  pillars of, 3, 5, 8, 51
Colombetti, Giovanna, xxii, 23, 76–78
Colombo, Matteo, 33–34, 45–46
Computation, 4–5, 7, 51, 56, 60, 237, 270n6
  contentless, 50–51
  and functionalism, 7–8
Computationalism, 3, 8–9, 56, 60
Conservative Enactive, Embodied account of Cognition (CEC), xv, 8, 10, 14, 16–17, 83, 256n6
Constraints-led approach, 22–24
Content
  based vs. involving, 91, 135, 262n3
  mysterianism, 49–50
  nonconceptual, 101
  nonpropositional, 100, 279n19
  perceptual, 147
  phenomenal, 11
  propositional, 10, 44, 107, 113
  representational, 11, 12, 14, 52, 55, 60–61, 71, 92, 98–100, 104, 135, 139, 147, 150–160, 166, 168–170, 178–184, 193, 195, 199–201
  truth conditional, 84, 110, 174, 201
  vs. vehicle (*see* Vehicle)
Content Involving account of Cognition (CIC), 14, 16, 130–133, 144, 177–178, 181–183, 198, 242–243
  unrestricted, 14, 19, 60, 88–89, 129–133, 172, 178, 182–183, 241, 243–244
Continuity
  evolutionary, 122, 128–137
  and functional discontinuity, 134
  psychological, 128–137
  skepticism, 136, 268n13
Correctness conditions, xii, xxi, 10–12, 100, 102–103, 112, 147, 152, 182–190, 193, 201, 264n4, 273nn3,5, 274n8
Crane, Tim, 11, 100, 133–134, 263n7
Craver, Carl, 4, 282n6
Csibra, Gergely, 268n10
Culture, xxii, 138, 139, 253
  and enculturation, 79–80, 232, 253
Currie, Gregory, 184, 208, 276n7

Dale, Rick, 250
Dalmatian dog, 172–173
Darwin, Charles, 106–107, 131, 133–134
Davids, Keith, 23, 118
Davidson, Donald, 271n13, 278n18
De Brigard, Felipe, 28, 31, 217–218, 222–228, 278n16, 279nn20,22, 280nn23,25

Deese-Roediger-McDermott paradigm, 225
Default Mode Network, 218
Degenaar, Jan, 34, 269n2, 272n1
De Jaegher, Hanne, 77
Demarcation, 15–18. *See also* Mark, of the cognitive
Dennett, Daniel, 139, 148, 258n7
de Pinedo, Manuel, 77–78, 88, 262n5
Deregowski, Jan, 175
de Vignemont, Fréderique, 5
de-Wit, Lee, 66, 237, 245, 281n2
Dietrich, Eric, 250
Dijksterhuis, Eduard, 32
Di Paolo, Ezequiel, 29, 76–78, 255n1
Dostrovsky, Jonathan, 239
Drayson, Zoe, 247
Dretske, Fred, 11, 42–43, 109
Dreyfus, Herbert, 2, 12, 35, 40, 43, 101–103, 264n10
Duplex account of mind and cognition, xii, xvi, xx–xxi, 91, 176, 203. *See also* Multi-storey story
Dynamical systems, 9, 22–25, 101, 251–252

Ecological
 Dynamics, xix, 22–25, 85–88
 psychology, 1, 9, 10, 22–25, 75, 82–88

Eich, Eric, 279n21
Eisenberg, Ann, 212
Eliminativism, xii, 50–52, 116–117, 121, 129. *See also* Really Radical Enactive, Embodied account of Cognition (RREC)
Embodied formats, 5
Emulators, 179–180, 271n7
Engel, Andreas, 36–37
Equal Partner Principle, xviii, 21–22
Error
 memory, 222, 226, 228, 279n22, 280n27
 minimization, 70, 153
 prediction, 58–63, 70, 153–154, 260n3
 representational, 45, 79, 133–134, 145, 156
Expectation, 59, 69–74, 164, 180, 228, 278n17
Explanation
 best, xvii, 34, 73, 179–180, 185, 198
 dynamical, 250–252
 mechanistic, 3, 245–246
Extended
 Functionalism, 7, 8, 22, 246, 282n7
 mind or cognition, xiv, 1, 6–9, 22, 40, 211, 246–247, 274n11, 281n3
Extensive mind or cognition, xxii, 9, 25, 245–247, 252–254, 274n11, 281n3

Feynman, Richard, xi, 244
Fictionalism, 46–47, 258n10, 259n11
Fivush, Robyn, 207, 210–214, 219, 229, 276nn5–6
Flanagan, Owen, 95–96
Fodor, Jerry, 42, 44, 47, 107, 126, 151, 192, 271n8
Foglia, Lucia, 2, 179, 196–201
French, Lauren, 226
Fridland, Ellen, 136
Friston, Karl, 57
Froese, Tom, 10, 76–78, 82, 255n1
Fromhoff, Fayne, 211

Gabbay, Pamela, 278n16
Gallese, Vittorio, 5
Gallistel, Randy, 110
Garry, Maryanne, 226
Gerrans, Philip, xiii, 65, 167–169, 171, 182–183, 217–219, 271nn11–12, 272nn14,16,17, 273n3, 278n19
Gibson, J. J., 22–23, 82–87
Gibson, Martha, 67
Gładziejewski, Paweł, 109, 152–159, 270–271n6, 271n7
Godfrey-Smith, Peter, 114, 240
Goldman, Alvin, 2, 5, 30
Graham, George, 167
Grid cells, 240
Grush, Rick, 179–180, 196–201, 271n7, 273n2, 275n12
Gunther, York, 101

Habermas, Tilmann, 207, 212, 214, 219
Haden, Catherine, 207, 210, 212
Hafting, Torkel, 239
Hard Problem of Content (HPC), xviii, xx, xxvii, 41–53, 69–70, 81, 122–124, 128, 155–156, 171, 244, 265n14
Harley, Keryn, 211–212
Harvey, Matthew, 129–130, 262n4
Haugeland, John, 43, 124, 127, 133
Head-direction cells, 240
Hekkert, Paul, 6
Henrich, Joseph, 175
Heras-Escribano, Manuel, 77–78, 88, 262n5
Herman, David, 208, 276
Herskovits, Melville, 175
Hidden cause, 67–68, 73, 260n5
History of interactions, 25, 32–33, 70, 105, 117, 165, 172–174, 268n11
Hoerl, Christopher, 207–208, 213, 215
Hohwy, Jacob, xiii, 57, 64–65, 70, 72–73
Holden, John, 255n1
Holyoak, Keith, 231–235
Hominin, 274n9
  brain, 195
  toolmaking, 193–196, 274nn9–10, 275n12
Horwich, Paul, 47
Horst, Steven, 4, 126

Index

Hristovski, Robert, 118
Hufendiek, Rebekka, 84
Hurley, Susan, 256n4
Hyperintellectualism, 152, 269n2

I-conception of mind, 4–5, 8, 51. *See also* Intellectualism; Internalism
Imagining
  as-possible, 184–187, 273n5
  as-present, 184–187
  basic, 181–189, 193–201
  contentless, xxi–xxii
  episodic memory , 189
  hybrid account of, xxi, 188–193, 201, 274nn6,8, 275n12
  judgement, 189–190
  sensory, xxi, 183–186, 188–194, 198–201, 274n6
Individualism, 4
Inertia, 32
Information, 57, 59, 65–66, 71, 83, 151, 281nn1–2
  about vs. for, 87–88
  contentful, xiv, xviii, 26–31, 35–36, 39, 43–44, 52, 55, 60, 62, 68–69, 165–170, 233–238, 241, 245–249
  contentless, 41, 86–88, 238, 243–245, 262n4 (*see also* Information, covariant)
  as control, 42
  covariant, xiii–xiv, 9, 29–31, 41–42, 125, 237–238, 243–245 (*see also* Information, contentless)
  encoding, 27–31, 227–228, 237
  flow, 59–60, 170, 260n3
  pick-up, xiv, 29, 84–87, 92, 139, 166–171, 238
  processing, xiv, xviii, xxii, 7, 22, 26–31, 35–37, 39, 42, 52, 55, 60, 83, 86–87, 92, 139, 166–171, 233–238, 245–249, 258n7, 265n16, 271nn11–12
  recovery, 222, 228
  retrieval, 27, 29
  semantic, 130, 238 (*see also* Information, contentful)
  sensitivity, xiv, 9, 71, 82, 86, 92, 138–139, 237–238, 244–245
  sensory, 27, 62–63, 154
  storage, xiv, 84, 233–236, 261n9
  as structural similarity (*see* Structural similarity)
Integration
  of contentless and contentful forms of perception, 171–176
  intelligible, 166–171
  intermodal or multimodal, 163–165, 236
  intramodal, 165, 236
Intellectualism, 3–4, 61, 81–82, 87, 139
Intellectualized enactivism, 61
Intentional directedness, 103, 113, 116, 118, 138. *See also* Target
Intensionality, 43
Intentionality, 93–107, 109, 112, 114–120

Intentionality (cont.)
  contentless, xix, 102–103, 127, 139, 143, 268n11. *See also* Ur-intentionality
  original, 98, 107
Internalism, 4, 247, 281n3
Intraub, Helene, 224
Intuition, xvii–xviii, 33–34, 62, 81, 97, 99, 145, 148–149, 178, 219, 221–222, 232

Jackson, Frank, 264n11
Jacob, Pierre, 99
Janssen, Steve, 224

Kamitani, Yukiyasu, 65, 164
Kandel, Eric, 26–29, 93, 239–240, 257n1
Kaplan, David, 4, 251–252
Kelso, J. A. Scott, 24
Kennett, Jeanette, 216–217
Khalidi, Muhammed, 14
Kirchhoff, Michael, 274n11
Kirmayer, Laurence, 254
Kiverstein, Julian, 86, 118–119, 141, 266nn19–20, 268n11
Know-how, 2, 28–29, 36, 82, 208. *See also* Skill
Knowledge, 8, 29, 61, 64, 67–68, 125, 142–143, 154–155, 218, 257n2
Korbak, Tomasz, 122, 131–132, 266n2, 267n7
Kriegel, Uriah, 41, 45

Langland-Hassan, Peter, 183–193, 273n5, 274nn6–8, 279n19
Language, 2, 62, 96–99, 112–113, 126–129, 132, 141–142, 145–146, 194, 209, 229, 255n1, 261n9, 262n4, 263n4, 266n1, 268n11, 277n10, 278n17
Lavelle, Jane Suilin, 141–142
Learning, 26–28, 34–35, 68, 106, 117–118, 139–140, 194, 209, 211, 213–214, 239, 255n1, 261n8, 268n10
Lindsay, D. Stephen, 225
Loftus, Elizabeth, 225–226, 279n22
Lungarella, Max, 83
Lupyan, Gary, 175

Malafouris, Lambros, 195–196, 233, 261n1
Mandik, Peter, 180
Map, xxii, 105, 156–160, 233, 240–241, 265n17, 269n1. *See also* Brain, models or maps in
Mark. *See also* Demarcation
  of intentionality, 101
  of the cognitive, 13–15, 76–77, 247–249, 252, 282n7
  of the mental, 94, 167–168
Marr, David, 151, 272n17
Material Engagement Theory (MET), 193–196
Matthen, Mohan, 34, 48, 257n3, 259n13
Maturana, Humberto, 40

Index

McCauley, Robert, 175
McDermott, Kathleen, 217, 225–226
McDermott Sales, Jessica, 211, 229
McDowell, John, 91, 135, 145, 259n12, 267n8
McGinn, Colin, 188
McIlwain, Doris, 2
McIsaac, Heather, 279n21
Mechanism, 4–7, 25, 59, 118, 139–140, 166, 168, 242, 245–252, 265n13, 266n11, 268–269n11, 270–271n6, 272n17, 282n6
Medina, José, 144, 269n13, 273n4, 279n19
Memory. *See also* Remembering how
  autobiographical, xxi–xxii, 203, 206–216, 219–231, 276nn4–5, 277n11, 278n17, 281n1
  basic, 29
  declarative, 205–207, 216, 235–236, 275n9
  distortions, 218, 222–226, 228, 230, 242, 277n13, 280n25, 27
  encoding (*see* Information, encoding)
  episodic, xxi–xxii, 189, 203, 215–223, 233, 235–236, 275n1
  procedural, xxi, 28, 204, 206, 235
  shared, 229
  short-term, 219–220
  storage (*see* Information, storage)
  traces, 27–28
Menary, Richard, 128, 137, 261n1, 263n3, 267n4
Mendonça, Dina, 129, 177–180, 272n1
Merritt, Michele, 90
Michaelian, Kourken, 206, 216–217, 233–236, 275nn1–2, 276n5, 279n19, 280n27, 281n1
Miłkowski, Marcin, 42–43, 257n3, 265n14
Miller, David, 224–225
Miller, Peggy, 277n9
Millikan, Ruth, 43, 80, 84–85 105 108, 111–114 119, 130, 133, 226, 261n2, 265nn15,17, 265–266n18, 267n5
Model, 58, 151
  generative, 61, 154
  inner, 59, 63, 197–201 (*see also* Brain, model or maps in)
  mechanistic, 251
  mental (*see* Model, inner)
  as used by scientists, 74
Mole, Christopher, 271n10
Mori, Kazua, 226
Moser, Edvard, 239
Moser, May-Britt, 239
Moyal-Sharrock, Danièle, 233–234
Muller, Hans, 97, 99–100, 102–103, 114, 174, 263n6
Müller-Lyer illusion, 148, 174–176

Multiple temporal and spatial scales, 58, 81, 116, 119
Multi-storey story, xii, 137
Murre Jaap, 224
Musty thinking, xvii, 169
Mysterianism about content. *See* Content, mysterianism

Nadel, Lynn, 239
Nanay, Bence, 164–165
Narrative, xxi, 129, 203, 206–216, 218–221, 225, 228–232
Naturalism, xvii–xxi, 15, 18–19, 30–31, 41, 45–46, 50, 99–100, 104–107, 116–117, 120, 123, 127–129, 136–138, 140, 146, 149–150, 169–170, 179, 192, 201, 236, 272n16, 274nn7–8, 278n18
　American, 9
　explanatory, xviii, 30, 35, 41, 123–124
　relaxed, 122, 124
　restrictive, 123–126, 266n3
Natural Origins of Content (NOC), 92, 120, 122, 125–128, 136–146, 254
Natural signs, 80, 141–143
Navigation
　in insects, 108, 110
　in rats, 240–245
Neisser, Ulric, 223
Nelson, Katherine, 207–216, 219–220, 276nn5–6, 276n8, 277nn9–10, 278n17, 281n1
Neter, John, 224
Neurodynamics, xii, 236–245
Newell, Karl, 25
Nichols, Shaun, 181–182, 192
Nicolopoulou, Ageliki, 277n12
Nigro, Georgia, 223
Noble, Jason, 77–78, 88, 262n5
Noë, Alva, 7–9, 16–17, 256nn3–4, 256n6, 266n20
Norms
　Bayesian (*see* Bayesian, norms)
　Biological, 41, 43, 76–77, 104–106, 108, 116–117, 133–134
　for memory, 210, 231
　public, 12, 134
　semantic, 100–101, 113, 271–272n13
　sociocultural, 79, 119–120, 133–134, 145–146, 231, 253
　transient, 117

Objectivation, 114–116
O'Brien, Gerard, 14, 42, 141, 157, 268n11
Occam's razor, xvi
Offline cognition, 156, 179–180, 240
　vs. online cognition, 178
O'Keefe, John, 239
Ólafsdóttir, Halla, 240, 245
Opie, John, 14, 42, 141, 157, 268n11
O'Regan, J. Kevin, 16, 24
Orlandi, Nico, 160–163, 270n3, 271nn9–10
Outfielder problem, 82–83

Papineau, David, 105, 130
Pattee, Howard, 132
Penn, Derek, 131–135
Perception, 5, 8–9, 11–12, 16–17, 22, 31, 33–35, 57, 59, 64–65, 69, 72, 83–84, 89, 100, 147, 177–186, 222, 224, 251, 255n1, 256n6, 256n8, 259n1, 260n6, 261n8, 271nn7,9
  contentless, 148–149
  direct, 83
  by frogs, 115
  and judgment, 171–176
  multimodal, 163–165
  principles of, 152–153, 160
  processes of, 150–153, 160–163
  products of, 160–163
Pettit, Philip, 264n11
Pezdek, Kathy, 278n16
Piccinini, Gualteri, 50–51, 257n4
Pietroski, Paul, 114, 133
Pillemer, David, 212
Place cells, 239–240, 245
Pluralism, 45–47, 177–178, 188–189
Positioning system, xxii, 239–241, 244–245, 250
Povinelli, David, 131–135
Pragmatism, 36–38, 47
  neo–, 124
Precision estimates, 61–62, 153
Predictive Processing account of Cognition (PPC), xiii–xiv, xx, 56–74, 75, 81–84, 151–160, 180, 218, 259n1, 260nn3–5, 261n10
Presence-in-absence, 179–180
Price, Huw, xiii, 119–120, 133–134, 243–244, 278n18
Prinz, Jesse, 256n8
Proper function, 105–109, 226–227, 267n5
Pulvermuller, Friedemann, 2
Putnam, Hilary, 48, 133, 248

Radically Enactive Account of Behavior (REB), 15–17, 141–144. *See also* Behavior; Behaviorism
Rakoczy, Hannes, 268n10
Ramsey, William, 16–18, 28, 270n6. *See also* Ramsey's Rule
Ramsey's Rule, 18, 178, 198–199
Ramstead, Maxwell, 254
Ravenscroft, Ian, 184
Realism, 47–50, 154, 265n16, 271–272n13. *See also* Antirealism
Really Radical Enactive, Embodied account of Cognition (RREC), 12, 129–130, 262n3
Reddy, Leila, 66, 185
Re-enactment, xxi, 13, 28, 184, 201, 205, 235–236, 273n4. *See also* Simulation
Reese, Elaine, 207, 210–212
Reference, 94, 103–104, 111, 116, 121–122, 199, 263n5, 266n6, 280n26
Reflex, 102, 262n4

Remembering how, 204–205. *See also* Memory
Representation-hunger, 161, 178, 272n1. *See also* Absent; Abstract
Rescorla, Michael, 50–51, 110, 151–153, 155–156, 270nn3–4
Resemblance, 42, 157–160, 192, 199
Reuse, xiv, 2, 5, 178, 185, 217, 272n1
Rey, Georges, 264n9, 269n1
Reyna, Valerie, 222
Richardson, Michael, 224
Richner, Elizabeth, 277n12
Rickles, Dean, 23
Rietveld, Erik, 10, 70, 86, 117–119, 141, 266nn19–20, 268n11
Riley, Michael, 255n1
Ritchie, Jack, 106
Robinson, John, 279n21
Roediger, Henry, 27–28, 225–226, 275n1
Rohde, Marieke, 77
Rosch, Eleanor, 9
Rosenberg, Alex, 28, 44, 50, 69, 93, 106–107, 113, 116–117, 125, 257n1, 261n9
Rowlands, Mark, 6, 199, 243, 264–265n12
Roy, Jean-Michel, 56, 93–94, 99, 103–104, 114–117, 263nn1,2,5, 264n12
Rule following, 112, 113, 133
Rupert, Robert, 248–250, 252

Saltationism, 129
Sánchez-García, Raúl, 25
Satne, Glenda, 12, 90, 92, 122, 124, 127, 269n15, 278n18
Scaffold, 12, 90, 122–123, 128, 131, 137, 145, 191–193, 210, 212–213, 216, 220, 267n8
Scaling Down Objection, 131–132, 267n7
Scaling-up objection, 91
Schachtel, Ernest, 212
Schacter, Daniel, 206, 217, 277n13, 278nn14–15
Schechtman, Marya, 277
Seager, William, 272n15
Searle, John, 96–98, 126, 188, 239, 263n4
Segall, Marshall, 175
Selection
  by consequences, 105, 115, 117
  for, 33, 106, 109
Self, 209–211, 214–217, 219, 228–232, 276–277n8, 277n11, 278n17, 278n21, 281n1
Self-evidencing, 72–73, 260n4
Self-organizing, 23–24, 60, 76, 260n4
Sellars, Wilfrid, 112, 265n17
Sense, 103–104
Sense making, 77–82, 116, 172–173
Serre, Thomas, 66, 185
Shams, Ladan, 165

Shapiro, Lawrence, 14–15, 33, 48, 52–53, 55, 89, 104, 126–128, 181, 256n7, 257n3
Shea, Nick, 109–110, 265n16
Shimojo, Shinsuke, 164
Siegel, Susana, 84
Silberstein, Michael, 250–251
Silverman, David, 8
Simulation, xxi, 182, 184–185, 197–199, 201, 203, 216–221, 233, 235–236, 269n12, 278n19. *See also* Re-enactment
Sinigaglia, Corrado, 2, 5
Skill, 1, 22, 24–25, 36, 101–102, 118–119, 194, 201, 207–209, 212, 215, 256n6, 266n20, 268n11, 269n15, 277n11. *See also* Know-how
Smith, Linda, 70, 251
Social Interactionist Theory (SIT), 207–222, 231–232, 276n5
Spaulding, Shannon, 107–109
Sporns, Olaf, 70, 83
Standing for, 80, 99, 161, 199
Standing in, 33, 116, 157–159, 161–162, 197, 199, 272n13
Stapleton, Mog, 78
Stephan, Klaas, 57
Sterelny, Kim, 18, 130–133, 267n5
Stern, David, 233–234
Steward, Helen, 9, 256n5
Stich, Stephen, 43, 50–51
Storehouse metaphor, 234, 257n2
Strawson, Galen, 41, 277n11

Structural similarity, 42. *See also* Resemblance
Subpersonal, 142–143, 168–170, 247–248, 259n12
Sutton, John, xiv–xv, 1–2, 205, 213, 220, 229–231, 233–234, 253, 276n5, 279n21, 280n27
Swanson, Karen, 279n21
Symbols, 1, 6, 12, 124, 237
  external, 126–127
  public, 134, 145–146
Szpunar, Karl, 217

Tallis, Raymond, 196
Target, 22, 43–44, 50–52, 76, 92, 102, 104–105, 110–111, 115–116, 139–140, 142–144, 159, 190, 266n19, 268n11, 269n1
Teleosemantics, xix, 43–44, 104–116, 130–131, 133, 138–139, 156, 267n5. *See also* Biosemantics
Teleosemiotics, 51–52, 87–88, 107, 114, 140, 237, 246
Thagard, Paul, 36
Thelen, Esther, 251
Thinking, xii–xiii, 12, 28, 89–92, 97, 119–121, 126–128, 130, 132, 134–136, 138, 145–146, 172, 174, 177–178, 181, 192, 196, 203, 217–218, 220–221, 255n2, 261n9, 262nn3–4, 265n17, 266n1, 274n7, 278nn17–18
Thompson, Charles, 224

Thompson, Evan, 9–10, 78
Tomasello. Michael, 268n10
Tong, Frank, 65
Tonneau, François, 110
Toolmaking. *See* Hominin, toolmaking
Tower of Hanoi puzzle, 181, 201
Travis, Charles, 10, 71
Triangulation, 141, 143–144
Truth conditions, 11, 43–45, 97, 100, 110–113, 264n8, 266n1, 267n5
Tsuchiya Naotsugu, 66, 185
Tulving, Endel, 206, 275n1

Ultra Conservative Embodied account of Cognition, 5–7, 21
Unification, 57, 82, 125, 165
Ur-intentionality, 92, 93, 95, 103–104, 107, 116, 118–119, 144, 246, 266n19, 268n11

van den Herik, Jasper, 129–130, 262n4
van Dijk, Ludger, 86–88
van Leeuwen, Cees, 6
Varela, Francesco, 9, 77
Vehicle, xviii, 5–7, 33, 35, 42, 51, 157, 269n1, 277n10. *See also* Symbols
as distinct from content, 37, 247, 258n7, 281n4
extended, 7
public, 208–209
Veissière, Samuel, 254

Veldeman, Johan, 6–7
Veridicality conditions, 11, 45, 147, 152, 155, 183, 279n19
Verstijnen, Ilse, 6
Villalobos, Mario, 116–117, 125

Wade, Kimberley, 225
Waksberg, Joseph, 224
Wang, Qi, 210
Warneken, Felix, 268n10
Watson, Jason, 217
Weber, Andreas, 77
Wheeler, Michael, 247–248, 281nn3–4, 282n7
White, Sheldon, 212
Williamson, Kellie, 205
Williamson, Timothy, 125
Withagen, Rob, 86–88
Wittgenstein, Ludwig, 74, 112–113, 147, 233–234

Yablo, Stephen, 186, 273n5

Zaman, Widaad, 229
Zednik, Carlos, 251
Zhao, Jiaying, 271n10
Ziemke, Tom, 255n1